P9-DFM-803

How to Think about Evolution
& Other Bible-Science Controversies

L. Duane Thurman

InterVarsity Press
Downers Grove
Illinois 60515

InterVarsity Press is the book-publishing division of Inter-Varsity Christian Fellowship, a student movement active on campus at hundreds of universities, colleges and schools of nursing. For information about local and regional activities, write IVCF, 233 Langdon St., Madison, WI 53703.

Distributed in Canada through InterVarsity Press, 1875 Leslie St., Unit 10, Don Mills, Ontario M3B 2M5, Canada.

Acknowledgment is made to the following for permission to reprint copyrighted material: Biblical quotations, unless otherwise indicated, are from the Revised Standard Version of the Bible copyrighted 1946, 1952, © 1971, 1973 and used by permission. From Implications of Evolution by G. A. Kerkut copyrighted 1960 by Pergamon Press, Ltd. Used by permission. From "Perspectives" by John Tyler Bonner in American Scientist, vol. 49, copyrighted 1961 and used by permission. From Evolution and Christian Thought Today, Edited by Russell L. Mixter, Copyright © William B. Eerdmans Publishing Company 1959. Used by permission. From Evolution, Second Edition by Jay M. Savage. Copyright © 1963, 1969 by Holt, Rinehart and Winston, Inc. Reprinted by permission of Holt, Rinehart and Winston.

ISBN 0-87784-701-0

Library of Congress Catalog Card Number: 78-13799

Printed in the United States of America

19 18 17 16 15 14 13 12 11 10 9 8 7 6 5
95 94 93 92 91 90 89 88 87 86 85 84

To all who are interested in a broad understanding and peaceful discussion of controversial issues.

1 **The Renewed Controversy** 17

Creation-Evolution Conflicts before Darwin 17
The Darwinian Conflict 19
The Scopes Trial and After 20
The California Textbook Controversy 22
Responses of the Scientific Community 23
Analysis of the Responses 32
Comparison with Former Controversies 33

2 **Acquiring Tools to Sharpen Your Thinking** 37

The Analytic Approach 38
Defining the Problem and the Terms 40
Recognizing Assumptions 44
Pseudoscience and Nonscience 48
Data and Interpretations 49
Common Reasons for Conflict 51
Two Major World Views 53

3 **Science and Its Methods** 59

The Methods of Science 60
Interpreting the Evidence 66
Treating New Evidence 69
Logic and Science 71
Faith and Science 76
The Limits of Science 78
Sources of Knowledge 82

4 The Factual Side: Microevolution 85

Analysis of Evolution as a Fact 86
Individual Variation and Microevolution 89
Origin of Species 93

5 The Theoretical Side: Macroevolution 97

Origin of the Universe 98
Origin of Life 99
Origins of Groups above the Species Level 100
Interpretation of Fossil Evidence 104
The Influence of Fossil Dating 106
Origin of the Human Race 107
Phylogenetic Trees 108
Why Not Creation? 112

6 Creation 115

Biblical Interpretation 116
Guidelines for Understanding the Bible 117
Time and Biblical Creation 119
Creation in Six Twenty-four-hour Days 120
Creation in Geologic Time 124

7 Your Approach to Controversy 127

The Renewed Controversy 128
Science, Evolution and Origins 128
The Scientific Evidence 131
Scripture, Creation and Origins 132
Maintaining Perspective 134
Guidelines for Dealing with Bible-Science Conflicts 134

Notes 139

Figures

Figure 1: The General Viewpoint of Theism *55*
Figure 2: Two Ways of Analyzing Evidence *56*
Figure 3: Three Types of Reasoning *73*
Figure 4: Projection beyond Existing Data *81*
Figure 5: Three Areas of Human Knowledge *82*
Figure 6: Two Types of Speciation *96*
Figure 7: Three Interpretations of Evidence of Origins *113*
Figure 8: The Relationships between Versions of Creation and Evolution *130*

Preface

This book is intended to furnish high-school and college students with a balanced approach to the origin of life which is missing from most introductory biology, anthropology, evolution and geology texts. It is brief enough to avoid a reading overload when used with regular assignments and yet complete enough to stand alone. Consequently the standard evolutionary accounts will not be presented in detail here. For those who desire additional information, references to many books on this subject are included.

By presenting several common theories and their underlying assumptions, I hope to educate rather than indoctrinate. I have tried to present the positive side of several positions and to purge error. I have also sought to avoid the confusing dogmatism so often present in writings on this topic.

This guide to the creation-evolution controversy aims to help you think, clarify the issue and make your own decisions. Some background information is given but no attempt is made to prove or disprove anything. I realize that in both science and religion many persons would prefer to have an authority tell them what to think or believe rather than to think things out for themselves. But we have good minds and the freedom of choice. We should use these wisely—not for the indiscriminate deposit of both jewels and junk. Consequently, this objective approach will require more effort on your part than books dedicated to a single perspective. I have included an examination of the roles of both faith and fact and have presented a realistic picture of science and scientists. In short, this is a primer for your investigation of the creation-evolution controversy.

The views of many specialists have been brought together here for your convenience. They are interesting insights not found in standard texts. The ample use of quotations reduces any bias that I might impart in a paraphrase of the material. Each reference is documented to allow you to check its context or pursue a particular aspect in more detail. This will allow you to familiarize yourself with some of the people involved in this dispute. After all, this controversy is mainly the product of people's interpretations, not scientific evidence or biblical texts. I hope you find this diversity of quotes from authorities and references helpful.

The approach used here was first tried in 1965 when college students and Christian parents of public school children came to me with problems about evolution and creation. At that time I had little personal interest because for two reasons I saw no conflict. First, I had just completed my Ph.D. thesis at the University of California at Berkeley. There I was seldom allowed to call something factual which was not adequately supported by objective evidence. Consequently, I would not accept anyone's work as scientific fact unless it was similarly supported. When reading about evolution and creation, I was usually able to separate the facts from the unsupported assertions, even though some authors claimed all were facts. Second, I had read widely in the evolutionary literature as part of my research in microevolution and had a more balanced view of evolution in general. The students had only one short, biased version of origins characteristic of most biology texts and were not able to distinguish between supported and unsupported "facts."

I began to provide some balance for students by referring to and quoting from many authorities not readily available to them. I eventually assembled these into a sixteen-page outline (accompanied by two to four lectures) and used it in that form for ten years. Students and parents increasingly requested a text which (1) was shorter than most books on evolution, (2) presented several possibilities rather than just one version of origins, (3) contained quotes from authorities in books not readily available to them and (4) was documented sufficiently to allow a reader to check on the context or to pursue some topics further.

Everyone brings a unique background and set of beliefs to any topic, especially one like evolution or creation. I have my biases too. As both a Christian and a scientist I believe that neither Christianity nor science should be distorted in an effort to defend a particular version of origins. I make no firm choice of any of the popular models for origins. This may be interpreted by some evolutionists and creationists as a lukewarm or wishy-washy position but it may be more in accord with known evidence than their positions. I have been both a nonscientist and a scientist as well as both a Christian and a non-Christian. These perspectives give certain advantages in understanding the current controversy.

I wish to thank the many persons who helped me with this book. The foundation of my thinking was built by Dr. John Davidson at the University of Nebraska and was continued by the graduate faculty at the University of California at Berkeley. My first four years of teaching were with Dr. Robert Frost who helped develop several of my ideas in our team-taught courses in principles of biology and origins. I especially thank my wife Joan for her encouragement, daughter Sonya for help in typing a rough draft and the staff of Oral Roberts University for their help with the illustrations and final typing. Valuable suggestions were offered by Mr. James Anderson, Dr. Richard Couch, Dr. Carl Hamilton, Dr. Laird Harris, Dr. Roger Hartman, Dr. Walter Hearn, Dr. Edward Nelson, Dr. John Nelson, Mr. Lynn Nichols, Dr. Robert Voight and Mr. Hale Whistler who reviewed the manuscript. However, I claim full responsibility for any errors and invite readers to bring them to my attention.

The Renewed Controversy

1

In the fall of 1969, the textbook committee of a State Board of Education unanimously passed a petition on the equal status of theories of creation and evolution that startled science educators throughout the nation. This was not in one of the so-called Bible Belt states considered by some to be unsophisticated in science; it was in California, one of the more progressive states in science and education. Controversy between creationists and evolutionists immediately flared and spread to other states. This book deals with this controversy and how to think your way through creation-evolution and similar Bible-science controversies. Because every current issue is best understood in the context of preceding events, we will begin by briefly examining the history of some creation-evolution conflicts.

Creation-Evolution Conflicts before Darwin

The first basic ideas of organic evolution, though vaguely expressed, were conceived by early Greek philosophers. Anaximander (611-547 B.C.) proposed one of the earliest theories of spontaneous generation of life from primordial fluid. Aristotle (384-322 B.C.) believed in a perfecting principle continually operating to improve the living world.[1] But these germinal ideas were not developed until much later. With their development came conflict.

Both Copernicus and Kepler faced opposition from the church when they espoused the theory of a heliocentric universe. The split between the church and certain scientists was widened by the burning of Giordano Bruno in 1600. Bruno has been called a martyr of science. It was largely thought that he died because of his belief in the Copernican astronomical system. In fact, however, the reason given for his condemnation and execution was his espousal of a heretical view of the Trinity.[2] Nevertheless, his death served to emphasize the theological implications of Copernicanism and set the church against the advance of science. Later, Galileo, Descartes and Newton each faced opposition from the church. Because church authorities believed they held the sacred deposit of truth, any ideas and observations contrary to their beliefs were strongly opposed. Often offenders had to justify themselves publicly.

This opposition inhibited the origin and expression of new ideas from the Middle Ages on into later centuries. Because some church authorities believed that the study of nature did not help prepare for the imminent end of the world, they became antagonistic to secular learning. They had no desire to investigate nature with an open mind. Uncritical belief in nature as interpreted by the church authorities was substituted for direct observation and research.[3] This substitution of authority for observation spawned much myth and error.

During the late 1700s and early 1800s, many theories of evolution were suggested. Carolus Linnaeus, the most eminent naturalist of his time, was well known for his belief that every known species was specially created in its present form and place. But he altered this belief in the fixity of the species and later postulated that the species originated from created genera. For this he was strongly criticized by both Catholics and Protestants.

As Linnaeus, Buffon and many others, including Erasmus Darwin (Charles Darwin's grandfather), continued to propose ideas on evolution, they were opposed by church authorities. But they were also paving the way for the biggest event of all, an event which opened a path for the acceptance of evolution.

The Darwinian Conflict

Charles Darwin had been quietly gathering supporting evidence on his theory of evolution by natural selection for twenty years, when he was prompted to make his theory public. Alfred Russell Wallace had independently arrived at the same conclusion and wrote of this to him in a letter. In 1858 their ideas were read in a joint paper entitled, "On the Tendency of the Species to Form Varieties, and on the Perpetuation of Varieties and Species by Natural Means of Selection" before the Linnaean Society in London in 1858.[4] Darwin then quickly published a short version of his work. His book, *The Origin of Species by Means of Natural Selection or The Preservation of Favored Races in the Struggle for Life,* appeared in 1859. For the first time an effective mechanism of change was suggested and supported by considerable evidence. At this time the public and scientists were ready to accept the theory. You should notice, though, that the titles of both the initial paper and the book spoke primarily of change within species, not the origin of major groups such as phyla or classes which is central in the current conflict.

Church authorities strongly opposed Darwin's book but they no longer had a large majority of the public backing them. Each side had many enthusiastic supporters. Unfortunately, many of the most vocal supporters on both sides were not sufficiently familiar with their own version, let alone the opposing theory, and often substituted emotion, ridicule and name calling for rational discussion. The controversy was active in England, France, Germany, the United States and elsewhere. Perhaps the most famous encounter during this period was the debate in 1860 between Thomas Huxley, an English biologist, and Bishop Samuel Wilberforce before the British Association for the Advancement of Science.

During the late 1800s and the early 1900s, much of the controversy continued but at a lower level of intensity and public visibility. The Roman Catholic Church was now saying that the Catholic faith did not prevent one from holding Darwinian theory and that the church was no more opposed to it than to the theories of Galileo or Copernicus. Many Protestant pastors and laymen, however, still strongly opposed the teaching of any form of evolution.

The Scopes Trial and After

The second major creation-evolution controversy included the famous Scopes trial held in 1925 at Dayton, Tennessee. A new law made it illegal to teach any theory that denied the biblical story of the divine creation of man in favor of a theory that man descended from lower animals. The American Civil Liberties Union (ACLU) had offered to back any teacher who wished to test the law. John T. Scopes, a young high-school teacher who neither understood nor taught evolution was encouraged to do so.[5]

The prominence of William Jennings Bryan, who headed the prosecution, and Clarence Darrow of the ACLU attracted the attention of the press and brought nationwide publicity. Bryan, a former editor, congressman, secretary of state and three-time unsuccessful candidate for the presidency, was an excellent orator and a strong fundamentalist. Although

Scopes was convicted, the Tennessee Supreme Court reversed the decision on technical grounds two years later.[6]

After the Scopes trial, publishers of most high-school textbooks understandably avoided evolution. Classroom teachers, however, could usually add any aspects of evolution which they thought appropriate to achieve a proper education. Yet a survey in 1942 revealed that fewer than one-half of the high-school biology teachers even mentioned evolution in their courses.[7] It is not known how much of this was because of the intimidating effects of the Scopes trial or how much was due to a greater concern for the practical, factual aspects (for example, population genetics) rather than the philosophical aspects of evolution.

After the launching of the Russian sputnik in 1957, a careful examination of United States science education was initiated in all fields. As a result, the federally funded National Science Foundation supported several study groups, including the Biological Science Curriculum Study (BSCS). BSCS introduced a series of texts in 1964. Each had a slightly different emphasis, but all included the origin of life and the evolution of the major types of organisms. This was regarded as atheistic material by many parents and school boards. It was opposed in Texas and other states.

BSCS was not the only source of unrest. The federal government also funded the production of MACOS (Mankind: A Course of Study) which was adopted by some public schools for their social science curricula in 1963. MACOS was a study of behavior, including the behavior of certain animals and of certain primitive peoples, which implied an evolutionary approach. MACOS started well but also ran into opposition in the early 1970s. Congress was asked to censor the curriculum. By 1975, sales had dropped seventy per cent. Although Congress did not censor MACOS, it stopped funds and suspended federal support of all science curriculum projects until all NSF educational programs were reviewed. An amendment was also passed which required all materials pioneered by NSF to be open to parental inspection.[8]

In spite of opposition to BSCS and MACOS, the opportunities to teach evolution increased. In 1968, the U.S. Supreme Court ruled that Arkansas's 1928 antievolution law was unconstitutional because it violated the first and fourteenth amendments. It was expected that Mississippi, the last state with antievolution laws, would soon repeal its law, which it did in December 1970.[9] Finally, forty-five years after the Scopes trial, all states were free of antievolution laws. However, there was now in many states a different kind of restriction of alternatives in teaching about origins. Evolutionary theory was usually the *sole explanation* given for origins and was often taught as fact rather than theory. This sparked a much larger conflict which began with the California textbook controversy.

The California Textbook Controversy

On November 13, 1969, the evolutionists' tranquility was shattered when Jean E. Sumrall, Nell J. Seagraves and Barbara M. Taylor petitioned the California State Board of Education regarding a framework submitted on October 10, 1969 by the state advisory committee. The petitioners were specifically concerned with the current restrictions relating to the public school teaching about the origin of life and man. Here is their proposal:

1. Special divine creation is not just a theistic belief but can also be explained as a scholarly and scientifically valid doctrine by the Creation Research Society. Therefore, it deserves equal status with other scientific explanations concerning the origin of man.
2. The current science framework presents an unbalanced philosophical approach, i.e., atheistic humanism. Such an unbalanced approach is prohibited by law.

3. Christian children have equal rights in tax-supported schools. Since it has been declared unconstitutional to teach religion in public schools, it is equally unconstitutional to teach atheism.[10]
4. "Dogmatism [should] be changed to conditional statements where speculation is offered as explanation for origins ... science [ought to] emphasize 'how' and not 'ultimate cause' for origins."[11]

This petition, if accepted, would end the teaching of evolution in California public schools as the sole explanation for the origin of man and would require consideration of alternate theories, including scientific creation. The petition was voted on by the board and passed unanimously.[12] This petition went a step beyond the board's unanimous approval in December 1963 of a policy that future state texts should refer to Darwinian evolution as an important scientific *theory* or *hypothesis*. It also encouraged teachers to teach it as a theory, not as a permanent truth.[13]

Responses of the Scientific Community

News of this action first attracted nationwide attention when it appeared in the March 1970 issue of *Bioscience*, the journal of the American Institute of Biological Sciences (AIBS). Dr. Elwood Ehrle, editor of the Office of Biological Education page, described the new guidelines and urged interested readers to share their ideas with their colleagues and the people mentioned in the article. They were encouraged to do something, such as write to a newspaper, instead of being a silent majority in the biological community.[14] Respond they did! More than a dozen letters were printed in four succeeding issues of *Bioscience*. They were followed by a three-page article, "Fundamentalist Scientists Oppose Darwinian Evolution," in the October 1970 issue. Though a brief sampling of the letters will be included here, you are encouraged to read all of them.

Very few of the letters to the editor published in *Bioscience* were either completely for or completely against the amended guidelines. Most approved of some aspects, such as the presentation of alternative theories and the reduction of scientific dogmatism. Even evolutionists who disagreed with several aspects of the religious bias of the petitioners respected the emphasis on openness.

This mixed feeling is probably best exemplified by Michael Byer whose objection was so strong as to "urge civil disobedience and mass protests" if the law would not allow the use of supplementary pamphlets to "present the objections to theories of special creation and spontaneous generation (and to be fair, those to evolution, if any). . . ."[15] Byer admitted, however, that "perhaps *we* have been guilty of too dogmatic a view for too long. By presenting each idea historically, with arguments pro and con, we could invite more student participation and interest. I am convinced that the truth, i.e., evolution, would thus become more forceful and convincing than ever in the end. I am almost certain that this was not the intention of the Board!"

The board's action left one biologist "absolutely horrified. . . . The fact that the belief of a minority of fundamentalist sects is allowed to enter the official state textbooks is also a violation, I would think, of the separation of church and state."[16]

Another writer shared this dismay at the California ruling but agreed that "science teaching by and large still contains very little argument, even though science itself *is* argument."[17] He also suggested that science students "may benefit from examining the presuppositions, evidence, and argument on which a scientific theory is based. . . . They might well benefit by pondering the relative merits of different kinds of presuppositions and evidence. . . . Perhaps we have been guilty of doing the snow job of which the creationists accuse us."

Others agreed to the presentation of alternative theories but objected to "elevating the Biblical theory of creation to the status of a scientific hypothesis rivaling Darwin's theory of evolution."[18]

Carl Gans was especially concerned about the apparent legislating of "the way each hypothesis may be explained."[19] Another evolutionary biologist was "disturbed" that this ruling "will so greatly influence the academic freedom of an entire nation by effectively forcing most publishers to treat any issue according to the specifications of the California State Board of Education."[20] She did not, however, object to presenting all theories on origins, literal genesis included, if done in such a way that the students could evaluate both pros and cons and make their own decisions. If handled "from an intellectual point of view and not an emotional one," she said, "I believe this is a much healthier and more scientific approach to the problem. Too often our students are told 'this is the way it is.' " Two other scientists also agreed that they and their colleagues had been too dogmatic.[21]

Some people described the new ruling as "frightening" and objected to giving equal weight to all theories because "the student is left with the idea that there is no scientific evidence which supports one theory over another. Science should be presented in a science class and religion in a class on religion (on Sunday)."[22]

Robert Tricardo thought it was incredible that "practicing scientists expect teenagers to decide for themselves the matter of the origin of life after having completed a one-year course in elementary biology."[23] He cited fundamentalist religious backgrounds in some students and intimidating evolutionist teachers as problems in objective decision making. He proposed that "instead of an 'accept now-understand later' philosophy with regard to teaching young people the origin of life, we might ask them to consider the facts, and to reserve judgment until their informational resources are substantial enough to permit intelligent decisions to be made."

Of the few printed letters which were very biased, Kraatz was the most dogmatic. After recounting studies which had convinced him of "the truth of this concept," he flatly stated, "There can be no other explanation!"[24]

He was amazed "that such closed minds . . . could exist in this Scientific Age." He also ridiculed the creationists by calling their actions an "ignorant attack on science and . . . disservice to science students" and "intellectual dishonesty that would deny the facts."

Another biased but more scientific respondent spoke of teachers who "often use evolution (wrongly!) in an endeavor to upset the Christian faith of their impressionable students."[25] He saw this new ruling as an advance because "it will compel teachers to be more scholarly in their treatment of the evolution topic. . . . The parents and State Board of Education in California have reacted against frequent unjustified anti-theological dogmatism and excess stress upon macroevolution as distinguished from diversification."

The series in *Bioscience* was soon followed by a different series in the *American Biology Teacher* (ABT), the journal of the National Association of Biology Teachers (NABT). The controversy in this journal was started in November 1970 by an article entitled "A Challenge to Neo-Darwinism," which appeared in the section called "The Devil's Advocate." This section was headed by the following statement from the editor:

> The right to dissent, to question and challenge the norm of popular ideas, is basic to a free society and the democratic process. If our science classes are to be centers of inquiry thinking, where we encourage students to express themselves freely, then we teachers must continually provide the image and leadership these students will seek and re-

> spect. Speak out—become the devil's advocate![26]

Dr. Duane Gish, who holds a Ph.D. in biochemistry from the University of California (Berkeley), issued a strong challenge to the evolutionists. Dr. Gish, one of the principal scientific creationists in this controversy, focused on the interpretation of the fossil evidence for the origin of major categories of organisms, such as families, orders, classes and phyla, which appear abruptly and without apparent transitional forms. He maintained that the regular and systematic gaps in the fossil record support creation theory, not evolution theory (which would be supported by fossil evidence of many, gradual transitions). Dr. Gish and other creationists accepted the idea of limited variation within basic types (that is, population change or microevolution). They also did not object to considering the theory of the evolution of major categories as a working hypothesis for those who prefer it.

> We do maintain, however, that special creation is more in accord with the critical evidence and offers a viable alternative to evolutionary theory. What we are pleading for is a *balanced* presentation in our schools, with a full disclosure of the evidence, regardless of which theory it favors. The dogmatic fashion in which evolution is usually taught in our schools and universities amounts to indoctrination and is as much the teaching of religion as if the theory of origins were restricted to the Book of Genesis.[27]

Unfortunately, in his challenge to the evolutionists, Dr. Gish made several apparent overstatements. His opponents responded to these instead of to the principal arguments of the challenge itself. Responses printed in the *American Biology Teacher* included two articles in January, one in February and two in May of 1971.

Also appearing in the May issue was a letter by Thomas Cleaver of BSCS who severely criticized the journal (ABT) for carrying an advertisement for *Biology: A Search for Order in Complexity*.[28] This high-school biology text, edited by

John N. Moore and Harold S. Slusher, was written by the textbook committee of the Creation Research Society. It was intended to provide at least one text offering both creation and evolution as alternative explanations for origins. Its strong antievolutionary and procreation biases, however, caused controversies in at least five states. The first case to reach the courts was in Indiana where the Marion County Superior Court ruled that the text violated the constitutional separation of church and state.[29]

The controversy became more visible when William Willoughby, a reporter for the *Evening Star and Daily News* in Washington, D.C., filed a complaint in U.S. District Court asking for termination of the "one-sided, biased, and damaging presentation" of evolution as the only explanation for origins. He especially objected to the BSCS series, which was funded by $7 million from the National Science Foundation and used in over forty per cent of U.S. school systems. "If there is any place in the world where there should be fair play, it should be in the scientific world. Yet the BSCS series deliberately excludes the argument for design in the origin of the universe from even a modicum of exposure."[30]

Science, the journal for the more than 100,000 members of the American Association for the Advancement of Science (AAAS), also became involved in the controversy, but on a smaller scale than *Bioscience* and the *American Biology Teacher.* Several letters to the editor, editorials and articles on the controversy were printed during 1972.[31]

The major theme of the 1972 national convention of the National Association of Biology Teachers was "Biology and Evolution." Among the talks featured in general sessions and minisymposia were: "Nothing in Biology Makes Sense Except in the Light of Evolution," Dobzhansky; "Ambivalent Aspects of Evolution," Hardin; "The Evolution of Design," Stebbins; "Evolution Is God's Method of Creation," Denack; "Evolution and the Law," Mayer; "Creation, Evolution, and the Historical Evidence," Gish; and "Evolution, Creation, and the Scientific Method," Moore. The last two talks by Gish and Moore were vigorously denounced by the moderator who used fifteen minutes of the question-and-answer session for his personal rebuttal. Gish and Moore were allowed to publish their replies to the moderator's comments. All of the talks given at the conference were published in the *American Biology Teacher.*[32]

The executive secretary of the NABT appealed to the Society for the Study of Evolution (SSE), the American Society of Naturalists and the Genetics Society of America for lists of professional biologists who were concerned to stop the inclusion of creationist theories in biology textbooks and were willing to be expert witnesses at court trials. They also requested from these organizations a brief list of issues involved in the teaching of evolution at the elementary and secondary levels, and responses to each of these issues. Their intent was to compile these responses to form a "briefing handbook." In the summer of 1977, NABT published a compendium which was intended to convince the public that creation should not be taught in the classroom. It included fourteen reprints of articles presenting legal opinions, official position papers and resolutions, and semitechnical arguments.[33]

Financial contributions were also sought for the "NABT Fund for the Freedom in Science Teaching" which was providing legal assistance to California teachers. A Tennessee law requiring alternate theories, including scientific creation, to be considered in teaching about origins had been passed in the spring of 1973. Although a University of Tennessee law professor volunteered to handle the case without fee, court costs and other fees had to be covered. Most of the previous contributions had been used to oppose the consideration of nonevolutionary alternatives to the question of origins in California.

In their appeal for support, the evolutionists claimed that a creationist success in California would lead to "extensive pressure to regulate the content of texts and the activity of teachers of every subject in the nation."[34] The idea of presenting more than one alternative on controversial problems, though a good scientific method, apparently disturbed some persons in the scientific community.

Other responses to the textbook controversy included the passing of official resolutions by the Academic Senate of the University of California, the American Association for the Advancement of Science and the National Academy of Science. These were published in the January 1973 issue of the *American Biology Teacher*. Additional resolutions were published in the *AIBS News* in 1973 and in the February 5, 1977 issue of *Science News*. This last resolution was a strong 650-word statement issued by the American Humanist Association, signed by "179 prominent scientists, educators and religious leaders," and published in the January/February *Humanist*, along with 18 pages of supporting articles.[35] This resolution was also being sent to major school districts, calling on them to oppose the "equal time" laws pending in several state legislatures. The statement drew many replies, mostly from creationists, but only two letters were published.

These actions by the scientific community were aimed at limiting legislative victories by creationists. When the new guidelines for selection of California texts were adopted, creationists in other states such as Alabama, Michigan, Indiana, Virginia, Wisconsin, Texas and Tennessee initiated actions to achieve similar goals. This often took the form of legislative proposals, some of which passed but later were declared unconstitutional. For example, the Tennessee law passed in 1973 was declared unconstitutional in 1975.[36]

Even California's new science guidelines were reconsidered and thrown out by state board action in January 1973.

This was preceded by a public hearing in November 1972 in which twenty-two scientists, including eleven creationists, testified concerning the inclusion of scientific creation in science texts. In the December board meeting, the Curriculum Development Commission recommended books for adoption, none of which included creation. The commission also passed two of the three resolutions made at the public hearing which were: "that dogmatism be changed to conditional statements where speculation is offered as explanation for origins," and "that science discuss 'how' and not 'ultimate cause' for origins."[37] Furthermore, a committee of seven, including two creationists, two theistic evolutionists and three members of the staff of the Department of Education, was appointed to edit the books which were eventually adopted.

In addition to the Creation Research Society, which became more widely known during the textbook controversy, the newly formed Institute for Creation Research, a division of Christian Heritage College, became quite active. In 1972, the ICR began publication of *Acts & Facts*, a small eight-page newsletter containing short articles on scientific creationism and news of the textbook controversies. It also included comments on the frequent debates on college campuses between scientific creationists, such as Dr. Gish and Dr. Henry Morris, and evolutionists. Creation-Life Publishers offer more than thirty books and booklets, mostly written by Gish or Morris, on various aspects of scientific creationism.

The ICR message, whether in debates, *Acts & Facts* or books, is basically the same: evolution is impossible, creation is as scientific as evolution for the origins of major groups, and either both or neither of the theories should be taught in public schools.

We shall see why this message is acceptable by some and unacceptable by others, including other scientists who are creationists.

Analysis of the Responses

Now that you know *what* happened, let us look at some reasons *why* this controversy was renewed.

It has been suggested that the creationists' endeavors were a reaction against political authority which reduced their power of choice and were an indication of frustration with the uncertainties of a technological society. Thus, they attacked science courses in an effort to "return to fundamentalist religion and traditional beliefs."[38] By ending the monopoly of evolution as the *only* answer offered to explain origins and having other theories included, creationists could see themselves as participating equals instead of powerless victims.

Actually, the creationists saw something that the scientists and science writers were either unaware of or were unwilling to admit. Science, which had generally been thought of as morally neutral and quite nonreligious, had entered the realm of religion in certain aspects of its teachings. Although the creationists knew this, they were initially unable to separate legitimate science from the new religion. The scientific creationists were, however, able to give the clarity needed to teach in a more scientific manner. The creationists rightly accepted the scientific aspects of evolution. But they maintained that when the limits of evidence were exceeded, the realm of hypotheses—in fact, *belief*—was entered. In this realm the creationists insisted that other possibilities should be considered. Exclusion of opposing ideas, especially when they interpret the same evidence in an equally satisfying manner, was considered to be more like religion than science.

If scientists examine all of the available data, evidence and alternatives before arriving at a conclusion, then why do they object so strongly to creation as an alternate model for the origin of the universe and life? Many scientists consider evolution a fact and creation only a myth. Surely a fact can withstand competition from a myth.

If, for example, scientists were asked to include in their texts the idea that the moon is made of Swiss cheese, would they respond as they did to the California ruling on evolution and creation? It would be an easy matter to consult specimens and make analyses of moon rocks and photographs and perhaps even arrange a personal visit to the moon to obtain more information to settle the matter conclusively. Therefore, this Swiss cheese theory should not be a threat to scientists. Creation, however, was sometimes responded to as though it were a threat. It is possible to say, then, that evolution is not founded on data as solid as the data on the composition of the moon.

Two clues suggest that some aspects of evolution are indeed a religion, at least as practiced by some evolutionists. First, we have already seen that some respondents were afraid to let the students decide for themselves. Second, the emotional tone of their responses to Gish's challenge in "The Devil's Advocate" column were more characteristic of a challenged political view or religion than science. New models and theories should be welcomed in science, though not in religion. It is good religious practice, but poor scientific method, to insist upon only one alternative. This is especially true when important aspects of a theory are so far beyond the reach of experimental and even observational science that they probably will never be known.

Comparison with Former Controversies

The California text book controversy was quite different from earlier conflicts. This action potentially affected every biology teacher in the nation. California constituted so large a portion of the text market (ten per cent) that publishers would probably have developed texts just for this state. Then they would have marketed these revised texts in all fifty states, because it is cheaper to produce one version than two.

California had set a precedent when many auto manufacturers put pollution-control equipment on cars, even though only California required it at the time. It was, therefore, feared that California's leadership would encourage other states to follow suit with textbooks including creation. Had this ruling occurred in a smaller state, the nation and the scientific community may not have been moved to such action.

This textbook controversy was not just another attempt to use the same old methods. It differed from the Scopes trial, with which it is commonly compared, in at least three significant points. First, unlike the Scopes trial which involved a restriction or limitation of material being taught (that is, evolution), this action did not result in a reduction of alternatives. In fact, it expanded the possibilities. This, of course, is sound scientific method and is much more defensible than restricting or omitting a topic.

Second, the controversy was not scientist vs. church authority, as in Darwin's case, or even scientist vs. fundamentalist layman, as in the Scopes trial. Instead, it was scientist vs. scientist, both agreeing on limited change (microevolution) but differing widely on large-scale change (macroevolution). The nonscientist petitioners were solidly backed by the Creation Research Society and its over four-hundred voting members, each with at least a master's degree and many with doctorates in some field of natural science. These scientists were actively supported by more than twelve hundred clergy, laymen and persons with only a bachelor's degree in science.

Third, this action had the strong public appeal of equality of opportunity and freedom from imposed restrictions—both highly prized American ideals. There were basically two ways to achieve equal consideration concerning the controversial aspects of evolution. The first attempt, that of presenting neither alternative, failed. The second approach, however, that of requiring that both versions of origins be considered, tried to regain the equality of opportunity lost when texts presented only one option (often as a fact instead of as a theory). Furthermore, the latter approach was proscientific, not antiscientific in nature as were the previous attempts.

This controversy also had some things in common with previous ones. The creationists still adhered to a literal interpretation of Genesis and considered the current evolutionary explanations of the origins of matter and life to be atheistic. The evolutionists again strongly objected to any reference to creation out of nothing as a possibility. They continued to extrapolate (project) much further beyond the data than the creationists would accept. Creationists did not have this much faith in the theory of evolution. Both groups still had their vocal, uninformed individuals with shallow understanding, high emotions and a tendency for overstatements. This, of course, clouded the issue and obscured a much larger, more rational silent majority of both evolutionists and creationists.

Acquiring Tools to Sharpen Your Thinking

2

The creation-evolution controversy is not between the Bible and science, as such. Instead, it is between some persons who hold *certain interpretations of the Bible* and other persons who maintain *certain extrapolations* or theories in science. (An extrapolation is an estimation beyond the known range of data. The estimation is assumed but not known to follow.) Dr. J. E. Orr, who has earned degrees in both science and theology, indicates that "the conflict appears to be between debatable theories of some scientists and doubtful interpretations of some expositors rather than a conflict between the facts of science and the text of scripture, between which there appears to be a very remarkable harmony."[1]

Your first step is to equip yourself to better understand this type of controversy. The aim of this chapter is to help you think critically about issues. Later chapters will introduce the scientific method and acquaint you with the various versions of creation and evolution. These intellectual tools should allow you to do some analytical thinking before making up your mind.

The Analytic Approach

Analytical thinking requires more time and effort than simply believing what someone else tells you. But it will bring you closer to the truth. In past centuries everyone, even the poets, philosophers and royalty, knew and discussed the latest scientific discoveries. Today even the most modern, well-educated people often do not. They believe that science is so complicated that years of study are necessary to understand it. It should be left to the experts, they feel, because while it is interesting, it is not relevant to problems in today's society. There are three major reasons why we should examine these assertions critically.

First, the creation-evolution controversy is still an issue today. And it will probably continue to be of public interest throughout your lifetime as a taxpayer and voter. A casual reading of current newspapers shows that many matters of public policy involve judgments about the future of scientific research, the limits of scientific expertise and the place of science in our society. If we are to decide about complicated issues such as cloning or combining different genetic codes to produce unfamiliar organisms, we must be able to think critically about scientific endeavors.

Second, learning to think critically will enable you to go beyond dogmatic assertions (which may sound like facts) to find the real answer. Dr. G. A. Kerkut, an English evolutionist, comments on the unwillingness to think analytically which occurs even among scientists:

> It is very depressing to find that many subjects are becoming encased in scientific dogmatism. The basic information is frequently overlooked or ignored and opinions become repeated so often and so loudly that they take on the

> tone of Laws. Although it does take a considerable amount of time, it is essential that the basic information is frequently re-examined and the conclusions analyzed. From time to time one must stop and attempt to think things out for oneself instead of just accepting the most widely quoted viewpoint.[2]

Two other scientists affirm this viewpoint in their discussion of the beginning of life:

> Certain hypotheses become acceptable and predictable mainly because they have gained familiarity.... Such ideas eventually change but may become almost ritual while they last: specialists basically listening at meetings to what all have already heard.[3]

One of the nation's prominent evolutionists, Dr. E. Mayr, also recognizes this scientific dogmatism when he wonders how many people realize that Darwin did not solve the problem indicated by the title of his book, *The Origin of Species.*[4] Darwin never demonstrated the formation of two or more species from a common source.

Creationists have already thought critically about evolution and have had some success in increasing the accuracy of statements on origins presented in public school biology texts, at least in California. The most vocal creationists, however, tend to overdo it by introducing their own religious dogmas which are not acceptable to many evolutionists and other creationists. Consequently, their valid criticisms are thrown out along with their dogmas. It seems that everyone would be better served if fewer dogmas were included in the discussion of scientific matters.

Third and finally, we have an obligation to think critically because that is an essential part of our humanity. All people, but especially Christians, are obligated to use the mental capacities they have been given. Some Christians have thought that this is not so. But more and more are recognizing that they have a duty to think even about matters of faith

—to face all of the facts "with intellectual humility, and with the very best reasoning ability that one can muster."[5] In the book *Your Mind Matters,* John Stott, a respected Christian clergyman, says that when we fail to use our minds, we descend to the level of animals.[6] In other words, we must use the minds which God has given us.

How do you handle controversial topics such as creation and evolution? Are you currently able to think through this controversy? Or do you give up and rely on your emotions to dictate your beliefs? Can you as a student, parent or teacher help others analyze conflicting positions? If you can master this controversy yourself, and make up your own mind, then you can help others do the same.

Analytical thinking is time-consuming and difficult. It would be much quicker to just believe what someone else tells you. Yet sound thinking, whether it is called science or merely common sense, is very useful. It will help you in many areas of your life such as consumer decisions, voting and evangelism.

Defining the Problem and the Terms

Unless you define the problem, how do you know where to look for an answer or recognize the answer if it comes? When the issue is clouded with emotion and hidden in irrelevancies, defining the problem is sometimes difficult, but always necessary.

Much of the creation-evolution conflict can be solved by defining such terms as *creation* and *evolution.* How many who use the term *creation* or *creationist* realize that there are at least six versions, each differing from each other and from evolution by various degrees? Different parties in this controversy often do not have the same version of creation in mind. Oftentimes they are each referring to a different concept.

Evolution probably causes even more confusion because

of some important differences in the way it is used. For example, Dr. J. M. Savage discusses both "the fact of evolution" and the "most generally accepted theory of evolution."[7] Is evolution a fact or is it a theory? Or both? It is not unusual for someone to consider factual what others consider theoretical. But why would the same person consider evolution both? The answer lies in the variable uses and definitions of this term. Other terms with similar variable uses and definitions such as *fact* and *theory* will be discussed in more detail in chapter three.

Evolution basically means an unrolling or a process of change in a certain direction. It is, however, the extent of change which is the main key to the controversy. As one of the major premises of his book, Dr. Ledyard Stebbins, one of the world's foremost plant evolutionists and author of several books on the subject, explains that evolution must be considered on three levels. Each level has a different dominant evolutionary process. The first level, *individual variation,* refers to the individual differences found within a single family or population. The second level, called *microevolution* (small change), is variation among different populations of the same species.[8] The variation among all the races of people is an example of this. The third level, called *macroevolution* (large change), refers to the separation and divergence of populations or population systems which eventually form different species, genera, families, orders and other major categories of organisms.

Still other levels of evolution, such as megaevolution or quantum evolution, are occasionally referred to by other evolutionists. But they fall within Stebbins's definition of macroevolution and will not be considered as separate levels here. Kerkut refers to the two lower levels as the "Special Theory of Evolution" and to major changes as the "General Theory of Evolution."[9] Others consider "evolution" as a process involving genetic change and differential reproduction. This is distinct from "evolutionary schemes" which describe the nature and magnitude of past changes in populations.[10]

Probably few, if any, creationists find any conflict between the first two levels of evolution and their interpretation of biblical creation. They fully accept the experimentally verifiable changes within a species (for example, antibiotic resistance in bacteria or the development of breeds of animals and varieties of plants). In fact, Dr. Walter Lammerts, editor of the books *Why Not Creation?* and *Scientific Studies in Creation,* is an active member of the Creation Research Society and a capable rose breeder. Most creationists, however, do not accept, and even actively oppose, macroevolution because of its direct conflict with their interpretations of the creation account in Genesis. This is the only level at which there is a serious conflict. And it is also the hardest level for evolutionists to defend because of its speculative nature. There is only indirect evidence to support it. Consequently, there is much room for differing interpretations between creationists and evolutionists.

Creationists commonly use the term *evolution* to refer only to macroevolution. For changes within the same species, they use the standard terminology of genetics and ecology, the two recognized disciplines studying these changes. The traditional evolutionist, however, uses the same term *evolution* to refer to all levels of change. This helps explain why Savage speaks of evolution as both a fact and a theory.[11] He can correctly write of evolution as a fact if he refers to demonstrated change within a species, that is, microevolution. He can also refer to evolution as a theory when considering the origins of major groups.

Confusion results for several reasons: (1) some evolutionists claim that evolution (meaning both microevolution and macroevolution) is a fact; (2) some creationists reject evolution because they cannot accept macroevolution; (3) some evolutionists interpret a rejection of evolution as a rejection of both microevolution and macroevolution. Obviously, the basic problem is that the two sides have not defined their terms. We must understand that the term *evolution* can mean either macroevolution or microevolution or both. When we are considering the creation-evolution problem, we must be clear as to which we mean.

The equivocation of terms has resulted in creationists being labeled antiscientific because of their refusal to accept supposedly sound experimental evidence (that is, microevolution). This problem has been apparent throughout the controversy and has been the factor which most confused the public.

By not defining their terms, evolutionists have implied that macroevolution (change by which man could evolve from a single cell) is as scientifically well established as microevolution. The creationists rightly object to this because there is a lack of direct, objective evidence to support macroevolution. Many evolutionists realize this but some apparently cannot resist the temptation to increase the certainty of a concept (macroevolution), in which they believe so strongly, by associating it with a more acceptable term (microevolution). Most evolutionists believe that the observations and experimental evidence for microevolution are adequate support for macroevolution. Most creationists, though, and some evolutionists lack this kind of faith.

Therefore, creationists need to understand that when evolutionists refer to evolution as a fact they are probably referring to microevolution only. Likewise, evolutionists should realize that when creationists reject evolution they are usually rejecting only macroevolution. If both groups would use the terms *macroevolution* and *microevolution* (or the special theory of evolution and the general theory), instead of the term *evolution*, the conflict would be largely alleviated.

Recognizing Assumptions

There is rarely such a thing as total objectivity, even in science. Everyone has some kind of bias, mental leaning, preconditioning and background which influences the kinds of things observed and the ways in which these observations are interpreted. The assumptions or presuppositions which someone brings to the facts, more than the facts themselves, often determine his or her conclusions. Therefore, whenever you read you should ask, "What are this person's underlying presuppositions?"[12]

According to Mayr, "The more different the backgrounds of two biologists the more different their attempts at causal explanation."[13] Louise Young adds that, "emotional responses to scientific theories are especially common in biology because these theories are involved in the areas that mean a great deal to us as human beings."[14] The scientist's personal enthusiasm for his own hypothesis often makes it difficult for him to be objective.

To what extent do these underlying human factors influence the kinds of evidence the scientist gathers and the conclusions formed from this data? Dr. Kerkut, an evolutionist and biochemist, emphasizes the importance of assumptions by devoting an entire chapter to them. (An assumption is something taken for granted, stated positively with great confidence but without objective proof.) Kerkut indicates that:

> Before one can decide that the theory of Evolution is the best explanation of the present-day range of forms of living material one should examine all the implications that such a theory may hold. Too often the theory is applied to, say, the development of the horse and then because it is held to be applicable there, it is extended to the rest of the animal kingdom with little or no further evidence.
>
> There are, however, seven basic assumptions that are often not mentioned during discussions of Evolution. Many evolutionists ignore the first six assumptions and

only consider the seventh. These are as follows:

(1) The first assumption is that non-living things gave rise to living material, i.e. spontaneous generation occurred.

(2) The second assumption is that spontaneous generation occurred only once.

The other assumptions all follow from the second one.

(3) The third assumption is that viruses, bacteria, plants and animals are all interrelated.

(4) The fourth assumption is that the Protozoa gave rise to the Metazoa.

(5) The fifth assumption is that the various invertebrate phyla are interrelated.

(6) The sixth assumption is that the invertebrates gave rise to the vertebrates.

(7) The seventh assumption is that within the vertebrates the fish gave rise to the amphibia, the amphibia to the reptiles, and the reptiles to the birds and mammals. Sometimes this is expressed in other words, i.e. that the modern amphibia and reptiles had a common ancestral stock, and so on.

For the initial purposes of the discussion on Evolution I shall consider that the supporters of the theory of Evolution hold that all these seven assumptions are valid, and that these assumptions form the "General Theory of Evolution."

The first point that I should like to make is that these seven assumptions by their nature are not capable of experimental verification. They assume that a certain series of events has occurred in the past. Thus though it may be possible to mimic some of these events under present-day conditions, this does not mean that these events must therefore have taken place in the past. All that it shows is that it is *possible* for such a change to take place. Thus to change a present-day reptile into a mammal, though of

> great interest, would not show the way in which the mammals did arise. Unfortunately we cannot bring about even this change; instead we have to depend upon limited circumstantial evidence for our assumptions, and it is now my intention to discuss the name of this evidence.[15]

One of the most important assumptions or beliefs underlying this conflict concerns the acceptance or rejection of a supernatural God at work in the origin of life. These different frames of reference or philosophical perspectives greatly influence the ways in which both creationists and evolutionists form conclusions from the same set of evidence. There are people on both sides of the fence (creationists and evolutionists) who believe in a personal Creator. The issue is not limited to a division between atheists and theists (those who believe in one God as Creator and Ruler of the universe). But, traditionally, creation has been thought of as the theistic position and evolution as atheistic.

People's assumptions can greatly influence the way they see things. It is quite possible for two people to see the same event and yet come to two very different understandings of what they saw. Police investigators try to get as many eyewitness reports of a crime as possible because each witness often reports the events in a different way.

Another interesting example which demonstrates the effects of one's preconditioning on perception involves placing one hand in hot water and the other hand in ice water. After holding your hands in the water for three minutes, suddenly place both hands in a third container of water at room temperature. The water at this intermediate temperature will feel warm to the cold hand and cool to the warm hand. Thus, you may perceive an event or set of data quite differently because of different preconditioning.

Kerkut also speaks of this kind of situation and how it relates to controversies in science:

> In many courses the student is obliged to read, assimilate

> and remember a vast amount of factual information on the quite false assumption that knowledge is the accumulation of facts. There seems so much to be learnt that those who come after him will have even more to learn, for more will be known. But this is not really so; much of what we learn today are only half truths or less and the students of tomorrow will not be bothered by many of the phlogistons that now torment our brains. . . .
>
> Often an incorrect idea or fact is accepted and takes the place of the correct one. An incorrect view can in this way successfully displace the correct view for many years and it requires very careful analysis and much experimental *data* to overthrow an accepted but incorrect theory. Most students become acquainted with many of the current concepts in biology whilst still at school and at an age when most people are, on the whole, uncritical. Then when they come to study the subject in more detail, they have in their minds several half truths and misconceptions which tend to prevent them from coming to a fresh appraisal of the situation. In addition, with a uniform pattern of education most students tend to have the same sort of educational background and so in conversation and discussion they accept common fallacies and agree on matters based on these fallacies.[16]

The tremendous accomplishments of science have led many people who are not familiar with the uncertainties of scientific research to believe that science gives us firm, correct answers to just about everything. Advertisers frequently show "doctors" in white coats and make liberal use of phrases such as, "It is a scientific fact that . . ." and "Scientific research has proven that . . ." to lend strength to a weak message. This is especially true concerning food and over-the-counter drugs such as aspirin and antacids. Some of these claims are true. But which ones? And how much of each is true?

Pseudoscience and Nonscience

To aid you to think critically, you should be able to distinguish sound scientific work from (1) "scientific" work improperly done (unscientific), (2) science falsely so-called (pseudoscience) and (3) philosophy and religion (nonscience). A careful study of chapter three should acquaint you with science and its methods, limitations and applicability. You will learn that science is not able nor designed to handle all realities, only those which are physical and objective. Yet, there are other important realities and other means of discovering truth, some of which are closely associated with and sometimes difficult to distinguish from science.

This means that you must listen carefully to all that the opposing side says to learn all you can before forming your conclusions. Give your opponents a fair hearing and the respect that you want them to give to your views. For the Christian, this is an opportunity to apply the golden rule (Mt. 7:12) or to "love your enemies" (Mt. 5:44; Lk. 6:27). Although you may wish to emphasize it differently, do not selectively ignore data. Be objective enough to see exactly what is and is not stated and implied by all sides.

There is nothing to fear from this comprehensive view. You will discover your own weak points as well as your opponent's strong points. This does not mean that you lose or that the other position is completely right. Neither of you may have the correct answer yet. Or both of you may each have a part of the truth, which approaches the whole truth more closely when combined. Dr. J. D. Thomas, professor of Bible, believes that any faith that constantly fears its destruction by new discoveries, that is, one that prospers in ignorance, is not satisfactory because it can be overthrown by mere enlightenment.[17]

It is not improper to hypothesize something which is later refuted. It is, however, improper to elevate an unsupported hypothesis to the status of a theory or fact without the necessary additional evidence. Concern should not be with winning an argument but with finding the truth. Personal satisfaction at the expense of scientific objectivity is hardly worth the price. Bartholemew and Lois Nagy summarize the proper approach when they say, "Such a search requires four ingredients: (a) meticulous, careful work; (b) patience; (c) consideration of findings by others, be it supportive or contradictory; and (d) courage, if necessary, to stand up against 'conventional wisdom.' "[18]

It is also very important to know what your opponent is not saying. Do not add to his or her statements. You may believe that they have stated or implied something which exists only in your mind. You may have misunderstood, or perhaps it is something you want them to say in order to make them look foolish. Both creationists and evolutionists have been guilty of using minority or former positions to represent the whole opposition. For example, some creationists insist that evolution is due completely to chance, a position which most evolutionists do not hold. They then proceed to disprove it mathematically. And evolutionists use the most extreme, least-known version of creation as representative of creationists in general.

Taking a part of a position to be representative of the whole is a logical fallacy. This means that it is an inappropriate method of arguing. It is no better than ridiculing your opponent's personal behavior or calling names. This sort of argumentation should not be used by scientists, Christians or anyone else who is concerned about getting at the truth.

Data and Interpretations

We have already learned that most of the conflict is between the various interpretations, not between the biblical text and basic scientific evidence. As Paul Little, a well-known evangelical writer, has said, "If we limit ourselves to what the

Bible actually says and to what the scientific facts actually are, we shrink the area of the conflict enormously."[19] Most of the conflict results from "well-meaning but misguided Christians" and scientists who make the Bible say more than it really says, and from scientific interpretations, which are distinct from the facts themselves.

What does the Bible say? Do you know the difference between the inspired Word and the interpreted Word? What kinds and amounts of scientific data exist? What do they mean? These are separate questions. The varied interpretations of the same biblical texts and scientific evidence have already given rise to at least four versions of evolution, six versions of creation and many religious denominations, each firmly convinced that its interpretation is the correct one.

Part of the variation in interpretation demonstrates the general rule that the less the Bible says, the more man (as interpreter) can say. For example, the ruination-reconstruction theory of creation is built upon what the Bible does not say in Genesis 1. The evolutionists have their own version of a gap theory to explain the lack of sufficient fossils to bridge the gaps among major groups of organisms. Many creationists claim these fossils do not exist. They are not satisfied with the evolutionists' reasons why the necessary fossils have not yet been found.

When the answer to a problem is not known, this should be recognized. Some problems do not yet have enough objective evidence for a substantial answer. Therefore, we must continue with our educated guesses (hypotheses) and beliefs until the necessary supporting evidence is discovered. Only then can we claim to know the answer.

It is intellectually respectable, in fact, a sign of maturity, to withhold a conclusion until sufficient evidence is available. J. T. Bonner is an evolutionist who recommends waiting: "The point, of course, is that although system may exist, we do not yet know what it is. But students are impatient and, at first, ignorance does not goad them into solving problems; they only see an unfinished canvas with great blotching holes in it. Just as the wrong solution to this difficult problem is easy, the right one is hard." He tells us not to feel that every problem has to have an answer right away. We should not be so impatient or biased that we allow "an hypothesis . . . to sneak into the false clothing of a fact."[20]

Some evolutionists attempt to awaken their colleagues to the fact that they have yet to discover sufficient evidence to support current evolutionary theory adequately. As long as scientists believe there is enough evidence, they will be less likely to discover it through scientific research. One concerned scientist has said:

> May I here humbly state as part of my biological *credo* that I believe that the theory of Evolution as presented by orthodox evolutionists is in many ways a satisfying explanation of some of the evidence. At the same time I think that the attempt to explain all living forms in terms of an evolution *from a unique source,* though a brave and valid attempt, is one that is premature and not satisfactorily supported by present-day evidence. It may in fact be shown ultimately to be the correct explanation, but the supporting evidence remains to be discovered. We can, if we like, believe that such an evolutionary system has taken place, but I for one do not think that "it has been proven beyond all reasonable doubt."[21]

I could cite or quote others but I think the point has been made. Do not be afraid to say, "I don't know," or, in the case of some problems, "I don't know but I will when the answer is discovered."

Common Reasons for Conflict

It may be apparent by now that there are several reasons why conflicts develop. Basically, these are:

1. *Insistence on a perfect fit between the Bible and current science.* Difficult questions arise when you attempt to reconcile a stable biblical text and a changing body of scientific knowledge. The best advice concerning certain points of conflict between science and the Bible is to admit the problem exists, avoid jumping to the conclusion that science or the Bible is wrong, and wait until new scientific discoveries clear up the problem as they have often done in the past. For example, archaeological discoveries have repeatedly cleared up apparent contradictions in the Bible. So Paul Little suggests that Christians should withhold judgment on certain controversies. But he also warns that "the Bible cannot be proven by archaeology."[22]

Do not force a harmony between science and Scripture where none exists. Give science a chance to move into the proper position. "A harmony of science and Scripture deemed most acceptable today might be completely outdated within thirty years, due chiefly to the endless revisions of scientific conclusions."[23] Science is a changing, self-correcting discipline, as you will soon see.

2. *Lack of sufficient information.* You must acquire the necessary additional information before forming your conclusions. Many conflicts could be avoided if people waited for more information.

3. *Misinformation.* Conflicts often arise because people are misinformed. This is particularly true in the sciences where the subject matter is especially complex. Because science advances so rapidly, it is difficult for textbook publishers to keep abreast of the latest developments. Sometimes an older, simpler but inaccurate theory is taught even after it has been disproven because it is easier for students to grasp.[24] At other times scientists themselves are misinformed. For example, in 1922 a fossil found in Nebraska was reported to be a primitive ape's tooth. It was later discovered to be only a peccary's tooth.[25] Some discoveries, such as the "Piltdown

man" were later found to be deliberate hoaxes.

4. *Misinterpretation.* Backgrounds and assumptions often influence the interpretation of scientific data.

In order to avoid controversy, an attempt must be made to separate the evidence for a theory from the interpretation of that evidence. This has not always been done with regard to the creation-evolution controversy. Because of this, conflict has arisen between those who interpret scientific data and those who interpret the Bible.

Misinterpretation has also arisen in this particular controversy because of what both sides believe is at stake. The most dogmatic creationists feel that faith in God and the authority of the Bible will be sacrificed if evolution is accepted. Most evolutionists, on the other hand, feel that admitting creation as a hypothesis for the origin of life is scientifically unacceptable. Their faith in rational, scientific method is being threatened. It would be helpful if creationists and evolutionists knew more about each other's views.

Two Major World Views

Each student of biology should be told of the different ways of viewing the world. Two of these ways, theism and naturalism, are relevant to the creation-evolution conflict and so should be clarified before meaningful discussion can begin. The table on page 54 compares the philosophical perspectives of theism and naturalism. Most, but not all, evolutionists will hold certain basic assumptions similar to those of naturalism. Most, but not all, theists will believe in creation. This table represents the two views most basic to the controversy.

These two world views are different ways of thinking about experience and different ways of framing initial questions about problems. The theist asks "why" and "who." Those who believe in naturalism, on the other hand, tend to frame questions in terms of "how." In science, questions

A comparison of two common philosophical perspectives used in interpreting the world and its origins.

Nontheistic Evolutionist	Creationist
Naturalism Nature is the sum total of reality. Knowledge of the world can be obtained entirely through the methods of science. There is no need to seek to explain the world in any other way.	**Theism** Natural science is not sufficient by itself to provide answers to life's questions (1 Cor. 2:12-14). Part of reality can be explained only in spiritual terms (Jn. 4:24). A complete view of reality recognizes both the natural and supernatural aspects.
Uniformity The Uniform Process Theory states that knowledge of the present is sufficient to explain the past and to predict the future. This is done on the basis of certain natural laws which are said to be changeless. There is no divine intervention in history.	**Sovereignty** The world was created by God (Heb. 11:3; Gen. 1:1; 2 Pet. 3:5-6). God does not change and has created a world which obeys certain uniform natural laws (Heb. 13:8; Prov. 3:19-20). But the history of the universe is not explicable in terms of natural laws alone because God is not bound by those laws (Mt. 19:26).
Chance (Causalism) Life began as the result of chance events. The end result of a chance event is a consequence rather than the achievement of a purpose. Because the present forms of life originated by chance, they could easily have arisen in some other form or not at all.	**Purpose (Teleology)** The world was created by God for his purpose. All history is a working out of God's plan (Col. 1:16-17; Eph. 1:9-12). Mankind was made in the image of God's own divine personality (Gen. 1:26; Rom. 8:28-29).

of "why" are often reframed as "how."

One common way in which many theists view the world and its relationship to God and man is shown in Figure 1. Theists believe that God reveals himself to mankind in two ways: his world (nature) and his Word. These two aspects of

Figure 1. The General Viewpoint of Theism

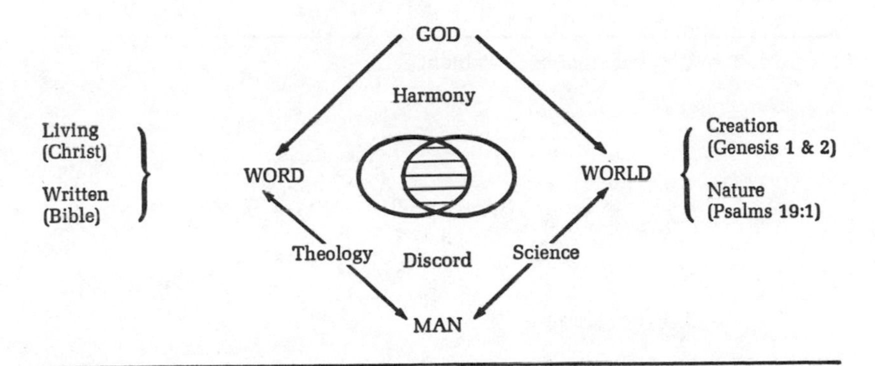

God's revelation are in harmony with each other. But this is not always evident to those who study the disciplines of science and theology. In some instances where the two aspects of God's revelation overlap (for example, the origin of matter, life and mankind), the systematic studies of science and theology indicate a conflict. As biblical interpretation becomes more refined and as science becomes more accurate, the apparent conflict between the two will disappear.

In contrast to this, those who believe in naturalism (which includes a large number of evolutionists) omit God's revelation through his Word as a valid approach to truth. They assert simply that a study of nature is the best way to discover reality.

Because they begin with different basic assumptions, those who believe in naturalism and those who believe in theism will look at the same evidence and come up with different conclusions. In many cases (but not *all*) theists will opt for creation as an explanation of the origin of life. Likewise, most people who espouse naturalism will choose evolution as the most acceptable explanation.

Figure 2 illustrates how this process occurs. Both views are organized systematically but are different because of differing assumptions, which are believed with different kinds of faith.

Figure 2. Two Ways of Analyzing Evidence

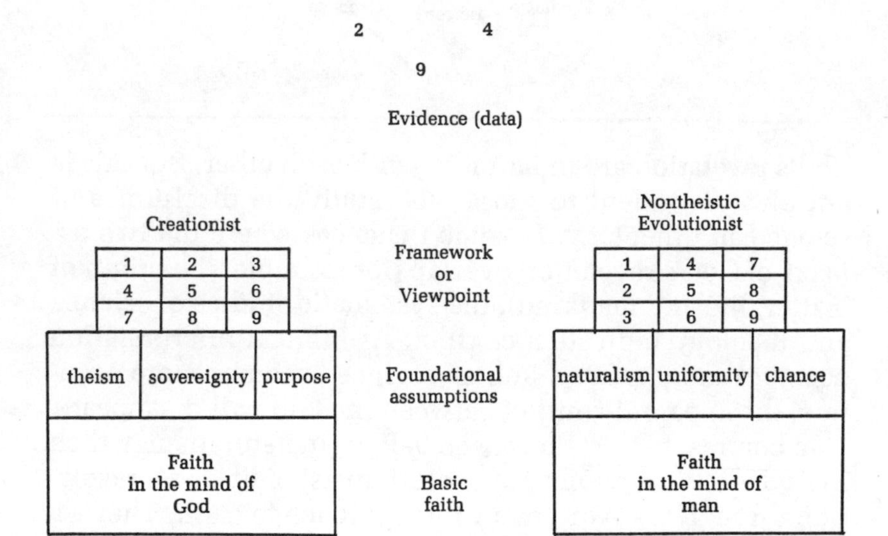

Given a set of randomly distributed numbers (1—9) which represent data, this common evidence can be interpreted and organized into at least two different arrangements, both logical and systematic. Each interpretation is based on a different set of assumptions (unprovable beliefs). These assumptions, in turn, rest on faith in two different sources: the mind of God as revealed in Scripture and creation; and the mind of man as revealed by his scientific study of nature. Nontheistic evolutionists restrict their interpretation to faith in the mind of man. Creationists have faith in both the mind of God and the mind of man, but they recognize that the mind of man is finite and fallen. The type of interpretation chosen for the data on origins depends more upon the assumptions and underlying faith of each person than on the common evidence.

Most creationists, when forming conclusions in areas of conflict between the two, will choose to rely upon faith in the mind of God because they believe that he has a better record; that is, he makes fewer errors than man does (presumably none). Most creationists accept valid scientific work, microevolution included, but reject most of the extrapolations based upon it. Both creationists and evolutionists reject each other's conclusions on the origins of life itself and of major kinds of organisms.

Science and Its Methods

3

The creation-evolution controversy involves differences among people. The evolutionists disagree with the creationists' claim that creation is just as scientific as macroevolution. But there is some question whether or not science can prove macroevolution and disprove creation, or vice versa. Just what can science do?

The important attainments of science have given it great stature and honor. Consequently, many laymen hold the unrealistic belief that if the scientific community says something it must be right. The complexities of modern science intimidate people into mental inaction and acceptance of whatever scientific authorities say. This encourages misuse of the words *science* and *scientific* to lend credibility to weak claims or opinions, especially in advertising. One way to develop a proper perspective on science and avoid its misuse is to know what science can and cannot do. That is the purpose of this chapter.

The Methods of Science

Science is characterized by its methods. If scientific methods cannot be properly applied, we are not dealing with science. Yet science is the most effective way to solve many problems, and so it should be used whenever possible. But we must first know which kinds of problems it can deal with legitimately. Once we understand scientific methods, we will also be able to detect the misapplication of science to problems it is not capable of answering. Such problems are nonscientific, as opposed to unscientific. The term *unscientific* applies to the use of incorrect methods on problems which science *is* capable of handling.

There are five steps in the scientific method as commonly taught and practiced. These are observation, statement of the problem, proposal of a hypothesis, gathering of data through observation and experimentation, and formation of a conclusion.

Observation. All science starts with observation by the five physical senses (taste, hearing, sight, touch and smell) or by extensions of these senses by telescopes, microscopes and other instruments. Anything which cannot be observed is outside the realm of science, that is, is nonscientific. Questions of the ultimate origins of our universe, life and major forms are actually nonscientific because they are events which nobody can observe. We may predict what *might have* happened but we will never know what *actually did* happen. This is an important point that gets confused in biology texts far too often.

All people use their physical senses to some degree but will usually see more or less than what is actually there. Correct observation is a difficult art, even for trained scientists. Error may creep in because of physical limitations such as color blindness, or preconceptions of what the observer wants, was told or thinks he ought to see.

Each observation must be also potentially or actually

repeatable by several other competent observers. This excludes observations which can be seen by only one person. Such observations would prevent verification by others, which is an essential part of science.[1]

How would this requirement of repeatable observations apply to questions of the origin of the universe and life? Technically, it excludes them. This does not, however, prevent scientists from proposing hypotheses (educated guesses) or speculations about what might have happened. This is a legitimate part of science, too. The point is that science cannot legitimately make any factual statement about what actually occurred, because no one observed it. But many text writers are either careless, ignorant or deceptive on this point.

The Question or Problem. A careful scientist asks questions about things most people take for granted. As observations raise questions in the scientist's mind, he tries to make as few assumptions as possible. Then he will ask questions of how, when and where a particular, problematic phenomenon occurs. These questions must be both relevant and testable.

Asking the right question is important in finding the right answer. Science cannot answer all kinds of questions, though. A question such as "Why does the universe exist?" and many other "why" questions cannot be tested and, therefore, are beyond the realm of science (that is, are nonscientific). There must be an objective way to test or verify possible answers to each question. Ideally, the problem will be reduced to two alternatives which can be answered by a simple yes or no.

Hypothesis: The Most Likely Answer. Now that a relevant, testable question has been asked about a repeatable observation, the next step is to propose a probable answer or educated guess called a hypothesis. Background can be a valuable help in suggesting this likely answer. But this same background can also close one's mind to consideration of the correct alternative, especially when this alternative is innovative or unusual.

At this point, the hypothesis is essentially an assertion (that is, a statement without evidence). Unfortunately, this provisional, tentative answer is sometimes promoted to the status of a theory or fact without acquiring additional supporting evidence or without undergoing sufficient testing.

Gathering Data: Obtaining Additional Observations. How does one know that a hypothesis is correct? A hypothesis is substantiated by gathering data to support it. This is a critical step overlooked by some who intuitively believe that their hypothesis is so reasonable and obviously correct that it needs no supporting evidence. A highly subjective "gut feeling" is helpful in making hypotheses but lacks validity as support for a conclusion in science.

Many things are logical but do not correspond with nature. For example, snakes have often been observed to eat other snakes, which naturally would disappear from sight while being swallowed. The logical consequences of two snakes simultaneously eating each other would be the disappearance of both. Simple observation of a test situation can easily establish the real outcome of this situation, which is different from that logically predicted. Sufficient observation or experimentation must be done to establish what actually happens—not what we want, or think might happen.

An appeal to other authorities is also not a valid way of supporting a hypothesis. We must question the basis of the authority's knowledge, to see if it is just opinion or if it has been scientifically established. A group of physicists at East Carolina University tried to duplicate Galileo's experiment in which he claimed to have dropped two objects, one heavier than the other, and found that both landed at the same time. The physicists filmed the descent of two balls of identical size but different weights in slow motion during a precision re-enactment. The heavier ball outdistanced the lighter one by twenty-five feet. They concluded that Galileo never actually performed the experiment as he claimed, but used it as an illustration.[2] Thus, relying only upon an appeal to Galileo as an authority to substantiate a hypothesis would have been a mistake.

Additional data may be gathered by further observation of the same type that sparked the initial question or by observation under controlled conditions. Every experiment should have a parallel test identical in all respects except for the variable factor being studied. This control group is necessary to distinguish the effects that would happen anyway. In other words, just having patients take a certain dosage and kind of medicine does not necessarily mean that their recovery was due to the medicine. Comparison with an identical group, which was given a placebo (medicine without the active ingredient) may show that as many, perhaps even more, would have recovered without any medicine.

It is not enough just to treat one group and not the other, for it is widely known that the knowledge of receiving treatment produces strong psychological effects, including cures. Claude Villee cites the example of a large group of students at a Midwestern university who took an extra-large, daily dose of vitamin C and caught 65% fewer colds than they did the previous winter. This was a statistically significant improvement. However, a second, similar group of students which took a pill of the same size, color, shape and taste (but containing no vitamin C) also had a significant reduction (63%) in colds compared to the previous year. Both groups thought they were getting vitamin C to help prevent colds. Without the no-vitamin-C control group, we would not have known that other factors, not vitamin C, were responsible for these significant reductions in the number of colds.[3]

This careful experimentation is often time-consuming, expensive and difficult to conduct without introducing bias.

Bias may arise in the observations, in the design of the experiment and in the amount and kind of data considered necessary for supporting or refuting the hypothesis. Yet, unless an adequate amount of representative data is obtained, false conclusions are likely.

The kind of evidence makes a great difference in the nature of the conclusion and its degree of certainty. At least five different kinds of evidence are encountered in the creation-evolution controversy.

The following example of *anecdotal* or *testimonial* evidence for running as a factor in preventing coronary heart disease was reported by Paul Milvy.[4] Clarence DeMar, a seventy-year-old marathon runner had coronary arteries with a cross-sectional area two to three times the average for men of his age. It is easy to assume that distance running, which is well known to increase arterial diameters, is responsible for this. But unless we know the arterial dimensions of his parents in order to tell whether or not he may have inherited larger arteries to begin with, we cannot say conclusively that running produced all of this change. In fact, he may have become a runner because his large arteries gave him the ability to run well, not vice versa.

Likewise, how do we know that the known limited changes observed in existing organisms are responsible for all of the differences required for the formation of major groups of organisms? The fact that these differences exist is not a reliable testimony that microevolution was necessarily the cause of macroevolution.

Positive evidence refers to observations that a predicted event happened. *Negative evidence* refers to the absence of evidence which would refute a hypothesis. That is, the fact that a theory has not yet been disproven is taken as evidence for its reliability. The statement that no factor except natural selection has ever influenced the course of evolution is considered by Krutch to be based on negative evidence.[5] He recognizes the very strong positive evidence for natural selection, but questions how we can be sure that it alone is sufficient to account for all the change. Just because natural selection is the only mechanism discovered at work thus far does not mean that other, less easily observable processes are not at work also. In other words, the nonexistence of alternative factors has not been conclusively demonstrated.

Direct evidence is an actual, demonstrated observation of an event or object. Direct evidence of an organism, such as a deer, would be confirmed by seeing the deer itself. *Indirect evidence* of the deer would be deerlike tracks and droppings, but not the deer itself. However, tracks of sheep, antelope or similar animals are more easily mistaken for evidence of the existence of a deer than the sighting of the animal would be.

Bernard Campbell, in his second edition of *Human Evolution,* gives his readers an idea of the kinds of evidence supporting human evolution:

> When we examine the evidence carefully, however, we realize that . . . the evidence is always indirect; inference after inference must be made as to the course of the evolution of man. The whole truth will perhaps never be known, but that does not negate the value of making a hypothetical interpretation of what evidence we can lay our hands on. . . .
>
> A first inference, which lies nearest to the facts, may be obtained from a study of fossil bones and teeth. The evidence that these ancient fragments furnish is not *directly* relevant because we cannot tell whether or not any fossil actually belonged to a human ancestor. Indeed, such a coincidence would be unlikely for any particular fossil individual, but whether it be the case or not can never be known for sure.[6]

His reference to the identity of fossil teeth probably stems from the earlier misidentification of a fossil peccary tooth as belonging to a manlike ape.

Interpreting the Evidence

Now that we have the evidence, what does it mean? The fifth step in the scientific method is the *formation of a conclusion* through interpreting the evidence. Can we keep human enthusiasm and bias out of this interpretation? Or will the investigator go overboard and become unscientific in his conclusions? It is very tempting for a limited amount of circumstantial evidence in favor of the hypothesis, which the investigator thought was the right one in the first place, to make the theory a "proven fact." Detachment and objectivity are required, especially in emotional topics such as the creation-evolution issue, to skillfully judge how strongly the evidence supports the hypothesis.

It is incorrect in biology to conclude that a hypothesis has ever been proven. "No scientific evidence, regardless of strength, can ever prove the absolute validity of any hypothesis. This inability is inherent in the nature of scientific methodology. . . . At best, we can only state that experimental evidence 'supports' or 'is consistent with' specific hypotheses."[7]

The results are never all in. There will always be room for more and better evidence, some even contradictory, which will change the certainty of the conclusion. An old story tells of the new social workers that were investigating the cause of drunkenness. They observed that all persons drinking many gin and sodas got drunk. The same was true for those who drank many rum and sodas or Scotch and sodas. Since soda water was the only common factor, they concluded that soda water was the cause of drunkenness. Even though this was a common factor, it was not the cause of the drunkenness. The real cause was the ethyl alcohol present in the rum, Scotch and gin.

Continued search and re-examination are necessary in science to purge it of error. New evidence may show that the hypothesis should be replaced by a better one, which is sometimes suggested by the nature of the new evidence. If no supporting evidence is found, it will remain an unsupported hypothesis.

Thomas Jefferson hypothesized that the westward exploration of North America would reveal mammoths and other prehistoric organisms which were then known only as fossils. No live mammoths have been found yet but the hypothesis is not necessarily disproven. Until every habitat has been examined and shown to be without mammoths, there is always the possibility that some exist to be found. An Argentine peccary, the coelacanth, dawn redwood and other organisms were known only as fossils until recently, when the first living specimens were found.

Spontaneous generation (the origin of life from nonliving materials) is another example of an unsupported but persistent hypothesis. It is a hypothesis essential to explain the origin of the life from which species evolved. Numerous attempts throughout the past few centuries have not yet demonstrated it. However, we may someday be able to produce life from chemical elements. But this still would not tell us how our present life actually originated. Meanwhile, spontaneous generation is assumed to be an established fact in far too many texts.

A hypothesis supported by a large body of different types of observations confirmed by many independent investigators may become a *theory*. A good theory: (1) explains or shows relationships among facts; (2) simplifies; (3) clarifies; (4) grows to relate additional facts, which means it is always tentative in scope; (5) predicts new facts and relationships; and (6) does not explain too much.

Both macroevolution and special creation have been criticized for explaining too much. Talbot points out that:

> Evolutionary theory has come under sharp criticisms for its lack of "falsifiability": there is no experimental finding which would be accepted by evolutionists as disproof of

> their theory. Put differently, an evolutionary "explanation" can be concocted for every possible finding. Such a theory, it seems, should be classed with what professional debunkers call fringe science. . . . *Every* fundamental assumption or set of assumptions, is difficult to falsify—especially where they deal with the unreachable past.[8]

Although the evolutionary trend is from simple to complex organisms, there are many exceptions, as in the cases of parasites and grass flowers. With regard to the distant past, much can be said without fear of contradiction because there is no direct evidence.

The special creationists do the same thing but in a different way, as Wallace indicates:

> We must concede that the extreme antievolution viewpoint cannot be refuted. He who is unalterably convinced at the outset that the different species of plants and animals did indeed arise through separate acts of special creation by an Intelligent Being will find nothing in this book compelling him to believe otherwise. . . .
>
> We merely conclude that it is unnecessary to invoke special creation as a mechanism to account for the diversity of life about us. Furthermore, we reject special creation as an adequate explanation because we can think of no means by which we can put it to a valid test, because we can imagine no observation falling outside the capabilities of a Creator possessing unlimited ability.[9]

More recently, the special creationists have countered the evolutionists' objections to calling creation a theory by now calling it a *model*. Like theories, models are not absolute truth, nor are they to be "worshipped as little gods of science."[10]

A *theory* is a generalization which is more certain than a hypothesis. Some theories, due to their high degree of certainty and wide acceptance by the scientific community, become *laws*. Yet even a law is not absolute truth. The nature of generalizations is well put by W. I. B. Beveridge:

> Generalisations can never be *proved*. They can be tested be seeing whether deductions made from them are in accord with experimental and observational facts, and if the results are not as predicted, the hypothesis or generalisation may be *disproved*. But a favourable result does not prove the generalisation, because the deduction from it may be true without its being true. Deductions, themselves correct, may be made from palpably absurd generalisations. . . . In strict logic, a generalisation is never proved and remains on probation indefinitely. . . .
>
> We should not place excessive trust in any generalisation, even widely accepted theories or laws. Newton did not regard the laws he formulated as the ultimate truth, but probably most following him did until Einstein showed how well-founded Newton's caution had been. In less fundamental matters how often do we see widely accepted notions superseded.[11]

The word *fact* needs some discussion because of its variable usage in this controversy. A scientific fact is an accurate discription of an object or event. The word is often used with a sense of absolute finality that establishes the point beyond question. Yet, *fact* is also used in place of *law*, *theory*, *hypothesis* and even *assertion*. Campbell illustrates this use of the term in his recent book, *Human Evolution:*

> Science does not claim to discover the final truth but only to put forward hypotheses based on the evidence that is available at the time of their presentation. Well-corroborated hypotheses are often treated as facts, and such a fact is that of organic evolution. . . . The evidence for evolution is overwhelming, and there is no known fact that either weakens the hypothesis or disproves it.[12]

Treating New Evidence

One desirable characteristic of a proper scientific attitude and a good theory is a willingness to accept new evidence and alternative explanations. Of course, this means that there is a chance of the existing (accepted) theory being contradicted and even displaced by a more appropriate theory. Some evolutionists encourage their readers to challenge evolution as part of a healthy scientific process, while others discourage any such attacks on their so-called sacred cows. Both of these attitudes are evident in the letters to the editors in response to the creationists' challenges in *Bioscience* and the *American Biology Teacher*.

Another interesting contrast in these attitudes is found in the prefaces of two books on evolution in the same "Modern Biology Series." Dr. Theodore Delevoryas, who specializes in fossil plants (paleobotany), demonstrates the ideal scientific attitude on this point.

> I hope it will be understood by the student that the evolutionary series presented throughout this book are not to be regarded as demonstrated facts. These series are, instead, logical conclusions held by many botanists based on observation and interpretation of the facts available to us. Additional information about plant structure, together with a reinterpretation of known facts, could lead to different hypotheses in the future. Evolutionary plant morphology is not a "closed book," and we can expect many new ideas as time goes on. It is quite possible—I sincerely hope probable—that readers will be stimulated toward re-evaluating some of the facts and ultimately presenting new ideas that may lead to a more accurate picture of change in the plant kingdom.[13]

Compare this approach with the following one which Savage presents to aspiring young biologists as they begin his book on animal evolution:

> No serious biologist today doubts the fact of evolution. . . . In this book we are not concerned with enumerating the so-called proofs of evolution. The fact of evolution is

> demonstrated on every side in all fields of biology and indeed forms the basic unifying principle in the study of living systems. We do not need to list evidences demonstrating the fact of evolution any more than we need to demonstrate the existence of mountain ranges. Rather, we will be concerned here with what is known about the process of evolution and with a survey of the several theories proposed to explain the process.[14]

Just why evolution should be considered science and yet not be subject to questioning is not clear. This attitude is usually characteristic of religion and other matters of belief.

Logic and Science

Logic is the science of correct reasoning. It is especially important to macroevolution because this level of change, which cannot be observed directly over eons of time, must be inferred from statements which are assumed to be true. Often the premise is an assumption which science either has not established or cannot ever establish as a fact, and therefore it must be believed as a matter of faith. Consequently, this neat, precise reasoning with valid, logical arguments can and does lead to conclusions which are not in accord with reality in nature.

Some scientists are keenly aware of these assumptions that are not capable of experimental or observational verification and must be taken on faith. They recognize that acceptance of a theory sometimes depends on whether or not they have the kind and amount of faith required to believe the underlying assumptions.

For example, there is no way to obtain direct evidence on how and when life began. Consequently, both evolutionists and creationists rely heavily upon assumptions concerning the origin of life. The creationists openly believe that God created life and its major forms, while the evolutionists believe that it happened by spontaneous generation and natural selection. Both rely on faith, although most of the evolutionists are reluctant to openly admit it. Their usual attitude is to assert that the creationists rely on faith, instead of sticking to the facts as they do. The faith of both is so strong that their respective assumptions are usually considered factual or unimportant.

The objective certainty of the assumptions does not affect the validity of the logical process. But if the certainty of the premises of a deductive argument is questioned, then the truth of the entire argument is questionable. Herbert Ross agrees that, "because our basic premises must be preceded by 'if,' all subsequent premises resting on this carry this 'if' with them."[15] Spontaneous generation is one of the problematic assumptions of evolutionary theory. It is required as a starting point for evolution. But spontaneous generation has never been demonstrated and the fact that it did occur *can* never be shown.

Both inductive and deductive reasoning are used in science. Each has its own use and neither is inferior to the other. They work together in *getting the facts* (induction) and *seeing what they mean* (deduction).

By inductive reasoning, a conclusion is drawn from premises that are direct reports of sense experience. Because such observations are limited by the extent of experience, inductive arguments are never absolutely certain. By induction, something can be said to be highly probable at best.

The statement "The sun will first appear in the east tomorrow" is a good example. It is based on the assumption that the sun always appears in the east every morning. That statement, however, is based on the inductive argument that the sun first appeared in the east yesterday and the day before and the day before that, so it must appear there every morning. However, no matter how many past observations we have made, we can never be certain about the future, so we can never have complete certainty in the statement that the sun will first appear in the east tomorrow.

Because the evidence for an inductive argument is never complete, an argument whose premises are based on induction will have only a probable conclusion. The conclusion can never be absolutely certain—it can only be as probable as its premise.

Figure 3 illustrates three basic types of reasoning used in evolution. Both a and b are the foundation of basic scientific findings (microevolution). Type c must be used for projecting hypotheses about origins.

Figure 3. Three Types of Reasoning

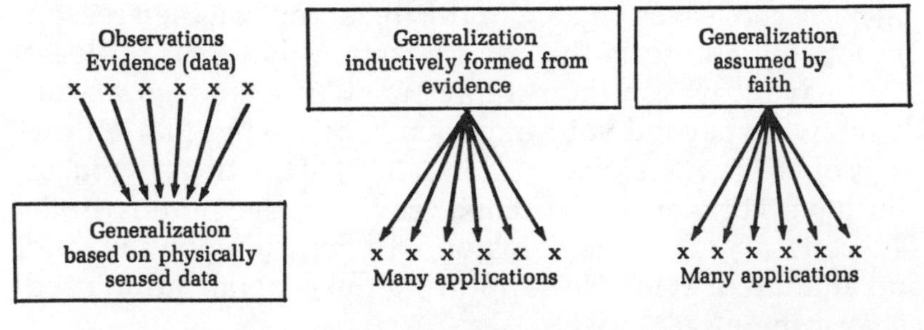

a. Inductive reasoning

b. Deductive reasoning from inductively derived generalizations

c. Deductive reasoning from generalizations based only on faith

In deductive reasoning, we can say that if the premises are true, and the argument proceeds logically, the conclusion *must* be true. This greater certainty, however, requires that the premises be true. Deductive arguments may begin either with premises based on sense experience or premises which are simply assumed to be true. If the premises are based on sense experience, they are as reliable as the inductive procedure from which they originated. If the premises are not based on sense experience, then they are based on faith.

According to logic texts, deductive reasoning in science is best done from an inductively derived generalization.

> If we are to have sound deductive arguments we must start with premises that ultimately have been justified by the method of induction. If we are to see the implications of the conclusions of our inductive arguments, we must construct valid deductive arguments. In the method of science we have a model for cooperation of deduction and induction—the scientist investigates, generalizes, and formulates hypotheses, but just as fundamental are his mathematical deductions of the consequences of his hypotheses, observations, and generalizations.[16]

An example of this process would be the observation of leaves on thousands of trees while on a summer trip. Having observed leaves on 95% of the trees, you can then reason inductively and form the generalization that trees (at least 95%) have leaves in the summertime. You can also reason deductively beyond your experience by saying that on the basis of your generalization, about 95% of the trees in places you have not visited also have leaves in the summertime. The certainty of your conclusion will depend upon the kind and amount of evidence supporting the generalization used as your premise.

Deductive reasoning may also be based on premises accepted in faith instead of on premises formed from verifiable, sensed observations. All conclusions based on assumed premises must be preceded by an "if." Unfortunately, human nature and enthusiasm often forget the "if" preceding the premises and faith converts them into facts.

J. B. Conant alerts us to the psychological factors in research:

> The stumbling way in which even the ablest of the scientists in every generation have had to fight through thickets of erroneous observations, misleading generalizations, inadequate formulations, and unconscious prejudice is

> rarely appreciated by those who obtain their scientific knowledge from textbooks. It is largely neglected by those expounders of the alleged scientific method who are fascinated by the logical rather than the psychological aspects of experimental investigations.[17]

The practice of objective, unbiased science is much more difficult than it appears from the sidelines. Types of improper reasoning are termed fallacies. Of the many types of fallacies evident in the current controversy, only two emotive types will be mentioned here.

Both creationists and evolutionists are guilty of the *ad populum* fallacy. Committing this fallacy means appealing to the common emotions of the multitudes. In their responses to the California textbook ruling, both creationists and evolutionists referred to the threat of communism in their appeals to the public.

Creationists were fighting for academic freedom—a value which they thought was necessary to avoid the kind of suppression scientists in some countries experience. Other creationists believe that communism developed from the theory of survival of the fittest and the idea that the population (group or party) is more important than the individual. It appears, however, that the basic ideas of communism were developed well before Darwin.

Likewise, some evolutionists likened the California textbook controversy to the political suppression of genetics by Lysenko, a ranking communist. Yet the creationists were requesting the expansion of ideas and opportunity, not restriction. Such emotional references blinded many readers to the real issues.

The other emotive fallacy is *ad hominem*, an attack on the person's character, intelligence or background. An example of this would be when evolutionists failed to respond to the basic creationist arguments and concentrated instead on how poorly they thought the arguments were presented.[18]

Emotive fallacies are very useful in political campaigns and salesmanship but should not be allowed to obscure scientific issues. You should refuse to be swept along by emotion, no matter how personally satisfying or infuriating you may find a particular position. Focus instead on the evidence and its interpretation. Emotional statements can cause the reader to forget or not notice that little or no evidence supports a particular argument.

Faith and Science

One point often mentioned by evolutionists is that evolution is based only on fact while creation is based only on faith. This implies that no factual evidence supports creation and no faith is involved in evolution. Neither of these implications is true. Both creationists and evolutionists rely on the same evidence as far as it goes. When considering topics for which there is no direct evidence (for example, the origin of life), both rely on faith, but of different kinds and degrees. Creationists openly admit their faith, something evolutionists are usually reluctant to do. But, as some evolutionists do admit, faith is used in evolution as well as in all scientific method.

Dr. Herbert Ross was mentioned earlier. He is the author of several books on evolution and was chairman of a symposium on "Logic in Biological Investigation." In his paper presented during the symposium, he said we "seem to imply that the scientific method is simply a construct of logic and reason, but this is not the case. The scientific method is actually a philosophy or faith espoused by certain persons having a particular aim, outlook, and practice."[19]

Evolutionists have a particularly difficult time regarding as factual *only* what is based on sense data (microevolution). As human beings they are influenced by what they are told to believe as students and by the desire to side with the most widely accepted evolutionary theory.

The basic role of faith in science as expressed by Dr. Loren Eiseley merits repeating here. Some evolutionists imply that similarity and order must be the result of evolution from a common ancestor. They assume that a universe governed by God would be unsystematic and erratic. But Eiseley says that the birth of science and the experimental method was due to

> *The sheer act of faith that the universe possessed order and could be interpreted by rational minds.* . . . The philosophy of experimental science . . . began its discoveries and made use of its method in the faith, not the knowledge, that it was dealing with a rational universe controlled by a Creator who did not act upon whim nor interfere with the forces He had set in operation. The experimental method succeeded beyond man's wildest dreams but the faith that brought it into being owes something to the Christian conception of the nature of God. It is surely one of the curious paradoxes of history that science, which professionally has little to do with faith, owes its origins to an act of faith that the universe can be rationally interpreted, and that science today is sustained by that assumption.[20]

Cordelia Barber asserts that the fossil record indicates the possibility of considerable descent with limited modification of the Genesis "kinds." She also discusses the role of faith in the interpretation of evidence of macroevolution:

> The chief objection to this view is that it is based on negative evidence, the absence of fossils. However, not even the most ardent evolutionary paleontologist anticipates that more than a few of the missing forms will ever be found. He crosses the gaps by faith in the principle of evolution because that seems a more realistic recourse to him than to invoke direct intervention from God. We can never be entirely certain just which gaps or discontinuities of record are real and which reflect the insufficiency of fossil collections. Likewise, one can never be dogmatic con-

> cerning just where the various groups of Genesis I may fit into modern classifications of plants and animals. Nevertheless, no apology needs to be made for this position.[21]

Many more examples are available but it should be clear by now that science is not all facts and no faith. A very respectable, reasonable faith is essential to scientific method itself. And an even greater faith is practiced with regard to the origin of the universe, life and its major forms (macroevolution). Such a belief may not, however, be openly acknowledged by evolutionists.

The Limits of Science

The enormous impact of science on modern society causes many people to erroneously regard it as potentially unlimited in its ability to solve problems. This is not necessarily the case. Science has its limits, as will be pointed out here.

First, we already know that science can answer only those questions which have testable hypotheses capable of support by sense data. Weisz and Keogh affirm that "science is useless as a tool to discover or evaluate any truth that cannot be tested experimentally."[22] A laboratory experiment cannot be devised to test the existence of God.

Science is also unable to make moral or value judgments, although scientists (as persons) are free to do so. Nor can it reveal the meaning in life. Dr. Warren Weaver characterizes science and its limits by saying that science is

> that amazingly successful, interesting, intriguing, elusive, and rewarding human process by means of which, within one particular framework of reference, men approach truth. This process moves in the direction of increasing precision and validity, but it does not reach perfection. It deals with certain very important aspects of experience—chiefly those that lend themselves to classification through quantitative regularities—but it excludes many

other important aspects of experience.[23]

Even this evaluation of science may be naive. Morris Goran, in his book *Science and Antiscience,* writes that "much of the scientific community equates science with certainty," which he calls a "myth."[24] He specifically mentions three points: (1) statements range from those which are accepted as established facts to those which are only educated guesses, but all of them are given to the student with equal emphasis; (2) data and concepts are packaged too neatly, hiding the loose ends; and (3) employment of the doctrine of the uniformity of nature is misleading because it is inherently uncertain.

One textbook writer, however, attempts to inform his readers that macroevolution cannot have the soundness that microevolution enjoys: "Admittedly, events that occurred in the past are not amenable to direct observation or experimental verification. There are no living eye-witnesses of very distant events. The process of evolution in the past has to be inferred."[25]

Ledyard Stebbins is quite frank in his orientation of the reader:

> I do not believe that anybody can quarrel with the point of view which maintains that all questions concerning the evolutionary history of organisms, in the absence of a significant fossil record, are pointless guessing games. The biologist who enters this field must resign himself to the fact that he can never achieve certainty....
>
> Hypotheses that embody these speculations are widely known and taught, and have acquired a credibility and illusion of certainty that is not justified by the facts and assumptions upon which they are based.[26]

Unfortunately, most readers either do not read Stebbins's preface or forget it because little of this openness and mature caution is evident in other writings on evolution. As Stebbins implies, it is quite appropriate to speculate but everyone must remember that these speculations are not facts to be memorized and quoted with more certainty than is justified. It is difficult to avoid making overly definite assertions which go beyond the data. In fact, it is the role of science to propose hypotheses to explain present data and go beyond these to predict future data. But these hypotheses must always be recognized as such—hypotheses rather than facts.

Within microevolution, scientists soon reach the limits of their data. They must then predict new observations and gather more data to verify their hypotheses. This works well where new data can be found by further research. But with regard to ultimate questions about origins, new data (fossils) are often unavailable or difficult to find and always inconclusive. Moore says that "most scientists recognize their limitations and the limitations of methods of scientific activity and do not press for first causes, as scientists. Most scholars will agree that first causes extend the searcher beyond the realm of scientific activity; first causes, then, become the primary concern of theologians, metaphysicians, and philosophers."[27]

Figure 4 illustrates several possible extrapolations (projections) beyond existing objective data (a). Macroevolution (beyond point x) lacks this objective data. Because results beyond the limits of experimentation or observation at point x vary widely, great caution is required when projecting possible curves beyond that point.

Each projection represents some known biological situation. In these situations further research produced the data which indicated that the extrapolations were correct. But direct evidence on how life and its major categories arose has not been found and probably cannot be. We may infer but will probably never know, with regard to origins, whether the data beyond point x resemble b, c, d, or still another curve. It will remain an idea held by faith. It is not a proven fact as is implied in school textbooks.

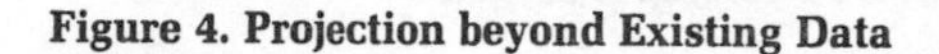

Figure 4. Projection beyond Existing Data

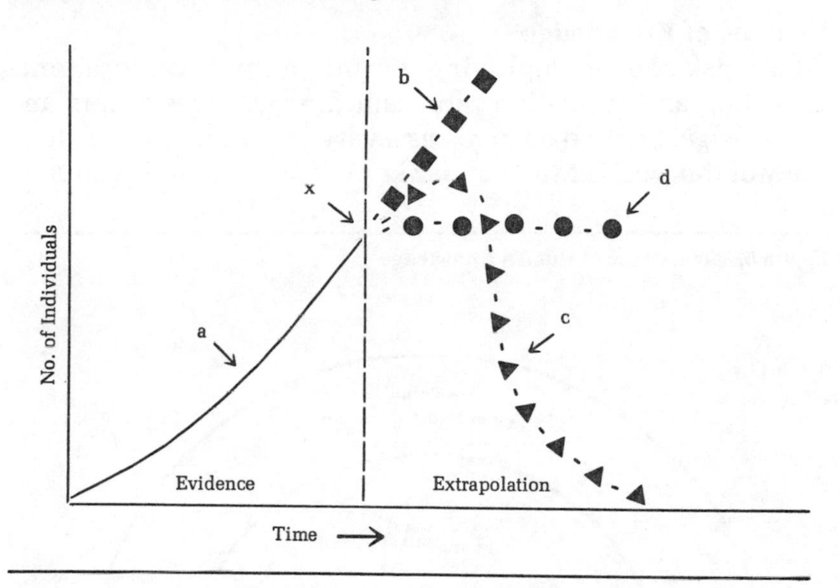

Dr. G. G. Simpson, a world-renowned paleontologist and author of more than a dozen books on evolution, recognizes the limits of science with regard to macroevolution:

> This is not to say that the whole mystery has been plumbed to its core or even that it ever will be. The ultimate mystery is beyond the reach of scientific investigation, and probably of the human mind. There is neither need nor excuse for postulation of nonmaterial intervention in the origin of life, the rise of man, or any other part of the long history of the material cosmos. Yet the origin of that cosmos and the causal principles of its history remain unexplained and inaccessible to science. Here is hidden the First Cause sought by theology and philosophy. The First Cause is not known and I suspect that it never will be known to living man. We may, if we are so inclined, worship it in our own

ways, but we certainly do not comprehend it.[28]

Sources of Knowledge

At the risk of oversimplifying, we might say that sensation, reflection and revelation represent three sources of human knowledge. Sense data gives us a very important but limited view of the world. Much of our knowledge (including much

Figure 5. Three Areas of Human Knowledge

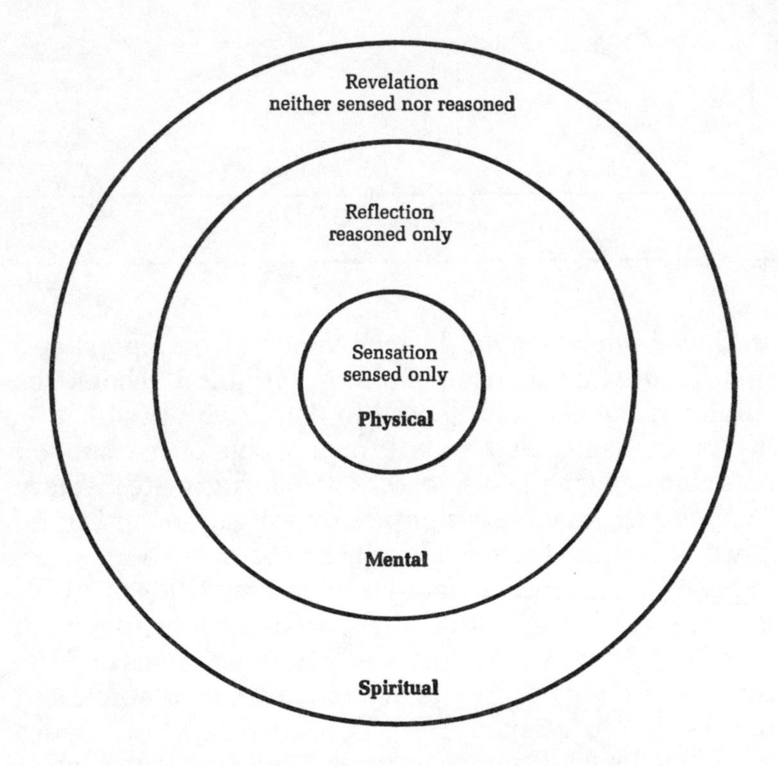

of science) originates in our own minds. Scientific theories, organizational structures, values and so on are all products of our mental processes (reflection). But we also need to unify our knowledge and this often comes through revelation. By this we mean direct revelation from God or revelation through the Bible, his written Word. Christians should keep in mind that true knowledge does not deny the worth of any of these aspects.

These three areas of knowledge are all interrelated. Our mental and spiritual preconceptions will influence how we interpret sense data. Sense data will influence our thoughts, feelings and spiritual experiences. These relationships are represented in Figure 5.

The most complete knowledge of reality requires a balance and integration of all three areas. Understanding human knowledge includes having a good grasp of science, its methods and limitations. If the diversity of human understanding can be grasped and if the limits of knowledge are recognized, the conflict between evolutionists and creationists may be substantially reduced.

The Factual Side: Microevolution

4

One of the most important causes of the creation-evolution controversy is the misunderstanding of words. The conflict is not *just* semantic, but this is one important factor. The word *evolution* causes much confusion because of its several meanings, implications and misuses. When someone asks, "Do you believe in evolution?" are they asking about the scientifically demonstrable changes within a species (microevolution), the theory that man evolved from molecules (macroevolution), or both?

In chapter two the word *evolution* was defined most basically as *change*. It was also pointed out that there is no conflict between creationists and evolutionists with regard to the two lower levels of evolution (that is, individual variation and microevolution). There is, however, a definite conflict with regard to macroevolution. And the conflict is enlarged by the evolutionists' use of a single term to refer to all levels of change.

Use of *evolution* for all levels of change implies that anyone who agrees that evolution is a fact (a statement which is appropriate for the two lower levels) is compelled to accept the theoretical level as factual also. When creationists and some evolutionists rightfully refuse to do this, they are labeled antiscientific for their refusal to accept "simple populational change." Evolutionists can understand why some people, including creationists, might not accept their particular theory of macroevolution, because not even all evolutionists agree at this level. Once the evolutionists understand that the creationist's rejection of evolution is at the theoretical level (macroevolution) and not at the experimentally validated levels (microevolution), emotions on both sides should subside enough that rational discussion can take place. I am convinced that when everyone uses and clearly understands the proper terminology, most of the reason for conflict will vanish.

The traditional evolutionary account will be omitted here because of its wide familiarity and availability in school texts. Instead, some important missing information will be added, often in the evolutionary specialists' own words as they "tell it like it is." This will add some balance to your previous instruction on origins, which was probably restricted to one version of evolution. Considering other existing theories is a more complex, but a more realistic, approach than is found in most texts.

Analysis of Evolution as a Fact

Having become acquainted with science, its methods and ways of thinking, let us analyze four statements on the factual nature of evolution.

Theodore Eaton says, "The fact of evolution has not been controversial among biologists during the present century at least, nor have serious scholars found need to question it in that time. Evolution as a real phenomenon is not a subject for debate, except occasionally among people who are either uninformed or for some reason anxious to impose dogma in the place of scientific learning."[1]

This statement should raise several questions in your mind. For example, was Eaton quoted out of context? (You have the specific reference to check on this for yourself.) Why have serious scholars in this century not examined the factualness of evolution? Why should evolution enjoy an immunity to re-examination? Should not a fact withstand examination much better than a theory or hypothesis? Why is anyone who asks questions (an essential part of the scientific method) considered unscholarly or dogmatic?

I think this example shows that even scientists are human, dogmatic and unscientific at times. Evolutionists are sometimes guilty of the same faults of which they accuse theologians and creationists. In spite of what some evolutionists say or imply when defending their beliefs, scientists are supposed to take little for granted. They are expected to question, rethink and repeat one another's works to verify them.

A second statement comes from Julian Huxley, one of the world's leading evolutionists who was invited to participate in the Darwin Centennial. He says that "evolution of life is no longer a theory; it is a fact and the basis of all our thinking. . . .We do not intend to get bogged down in semantics and definitions."[2]

In considering this quote, several questions might immediately occur to you. How could a clear definition of evolution (which would let his audience know exactly what he meant by this indefinite term) bog down a discussion? If Huxley had a precise definition, why was he unwilling to state it? Was there something to be gained by ambiguity? Would lumping the factual evidence with the speculative aspects of evolution result in greater acceptance of the whole package? Or was he afraid that only microevolution would be accepted and that macroevolutionary theory might be attacked because of its lack of direct evidence?

Theodosius Dobzhansky also comes on strongly with this statement about interpreting the evidence:

> Evolution as a historical fact was proved beyond reasonable doubt not later than in the closing decades of the 19th century. No one who takes the trouble to become familiar with the pertinent evidence has at present a valid reason to disbelieve that the living world, including man, is a product of an evolutionary development. . . . To be sure, anti-evolutionists still exist. But it is fair to say that most of them are not well informed, while the informed display biases which make arguments futile and facts useless. . . . Any such historical fact must be inferred on the basis of evidence which can be seen today. It is no mere matter of taste whether one rejects or accepts the inference. A hypothesis may, at a certain level of scientific knowledge, be forced upon every reasonable person.[3]

What can be done with this? Evolution is stated both as a historical fact and as a hypothesis. And if you disbelieve, you are either uninformed or biased. But take another look at Dobzhansky's remarks. If "historical fact" is taken to refer to microevolution and "hypothesis" is taken to refer to macroevolution, a proper perspective is established. You must remember that scientists are human and naturally become very enthusiastic about the product they have developed and are selling to you.

A fourth evolutionist and author of a freshman college biology text states that when biologists say they believe in evolution as a fact, they are referring to the concept of a biological world which is not static and unchanging, but shows evidence of changes in gene frequencies in species from one generation to the next.[4]

This appears to be a fair, accurate statement about microevolution. It does not ask you to believe unsupported assertions about macroevolution.

Individual Variation and Microevolution

The factual aspects of evolution are usually treated adequately in texts so will receive little further comment. However, we need to examine some species concepts for a perspective often omitted.

Species Concepts. Early species concepts were very nar row and unyielding. This sparked some of the earliest conflicts between scientists and creationists. Carolus Linnaeus, the foremost authority on classification in the mid 1700s, was a special creationist. He believed that the *kinds* of living things which God created (Gen. 1:11-28) were the current species of plants and animals in their present forms, which did not change with time. He also thought that no species became extinct and no new ones had been added since creation.[5] One commentator has said that, "indirectly, Linnaeus did as much to prepare the ground for a theory of evolution as if he had proposed such a theory himself."[6] His inflexible concept of the created species forced scientists to reject creation and look elsewhere for an explanation of origins. Later, after recognizing the existence of varieties and extinct organisms, Linnaeus omitted this strict definition from the last edition of his monumental book, *Systema Naturae.*[7]

The idea of the fixity of the species had, however, already been firmly implanted in the minds of others and persisted for more than a hundred years. Louis Agassiz, one of its active defenders, claimed that species were in their original proportions and localities.[8]

Many people thought of the Genesis kinds as representing species. The word *kind* can refer to species or to genera, families, orders, classes or phyla. Barber emphasizes the difference:

> In connection with the phrase "after its kind" we have emphasized that this cannot refer to species as we regard hem.... We have no way of knowing how much variation was to be inherent in each "kind." Since the fossil

> record does contain profound and persistent gaps between otherwise reasonably complete sequences of forms, it is an easy step to equate the genetic boundaries of the "kinds" with these gaps or lack of transitional forms between groups. Scripture fully allows what fossils seemingly indicate, namely, considerable descent with modification but always within predetermined limits.[9]

A species is a basic unit of classification which can be recognized and placed in a classification system without even considering evolution. The concept originated independent of evolutionary theory. It is defined in several different ways, depending upon the backgrounds and purposes of the investigators. Some, such as Linnaeus, defined a species as a group of individuals which looked alike. He said that there was an unbridged gap (that is, a lack of common characteristics) between species.[10] Mayr defines it genetically as "groups of interbreeding populations that are reproductively isolated from other such groups."[11]

Some say that a species is just a concept, the product of each scientist's judgment, while others say a species is a reality of nature.[12] This wide range of opinion on definitions might be summarized by saying that "it is increasingly clear that no single definition of species can be devised to express all the actual meanings of the word."[13] A species may be characterized in several different ways, but most definitions agree on the following: a species has certain designated characteristics in common; it usually does not interbreed with other species in nature; and if members of one species do breed with members of another, they usually will not produce fertile offspring.

Some situations in nature are so complex that it is largely a matter of personal judgment whether to consider the populations as one species with several varieties or as one genus with several species. In any case, it is important to determine what any author's definition of species is, especially when he or she is writing about the formation of new species.

Mutations. Every organism contains genes, some of which are mutant (changed) genes. Although most mutations are disadvantageous to the individual in its present environment, these same mutations may be neutral or even advantageous in another environment. If the environment changes, a mutation may allow an organism to adapt which would otherwise not have survived.

Mutations produce widely varying kinds and amounts of change. Some changes are so drastic that the organism dies during the early stages of development. Others will survive but die early because of severe deformity or malfunction. Consequently, most mutations available for study consist of small effects (micromutations) which are not serious handicaps to the development and survival of the individual. This is the commonest type of mutation referred to by both evolutionists and creationists.

There are, however, a few mutations with large effects (macromutations) available for study. These include flies with four wings instead of two, plants with petals and stamens typical of those in another genus and liverworts resembling another genus.[14] Although most evolutionists insist that such major changes are rare, R. B. Goldschmidt and others believe that macromutations are a better explanation for the sudden appearance of the many major groups which have no apparent transitions in the fossil record. Even though rare, with millions of years available, there need not be many macromutations to account for existing groups.

Recombination and Natural Selection. Interbreeding among different individuals within a population recombines mutations with different sets of genes. Some of these recombinations equip certain individuals to cope with their environment more successfully than others. Consequently, these better-adapted individuals tend to leave more offspring (with the favorable sets of genes) than the less-adapted types.

As the environment changes, and more of the better-adapted individuals survive, the population becomes slightly changed in the next generation. Succeeding generations produce increasing numbers of the better-adapted, mutant organisms. This process, called *natural selection*, produces races, varieties (plants), breeds (animals) and subspecies of organisms which exhibit slight but definite differences in appearance and function. Yet these various types can successfully interbreed with one another.

Mutations are random with respect to the environment and the needs of the individual organism. But as Louise Young indicates, natural selection is an "anti-chance agency making adaptive sense out of the chaos of countless combinations of mutant genes."[15] Thus, microevolution does not operate by blind chance as some creationists claim, although chance is definitely involved. For instance, the acorn which could produce the best-adapted oak tree might be eaten by a squirrel or crushed in the street by a car.

Although natural selection is very widely accepted, some major evolutionists caution that it is "in many instances hardly more than a postulate" whose "application raises numerous questions in almost every concrete case."[16] Ludwig von Bertalanffy points out that "if higher organization means selective advantage, the higher organization should have supplanted the lower ones. Every cross-section of nature, however, shows the most diverse levels of organization from unicellulars up to vertebrates, all maintaining themselves perfectly, and indeed all necessary for the maintenance of the biocoenosis."[17] Another scientist asks how evolutionists can be so sure that natural selection is the only mechanism, just because it is the only one observed thus far. He believes that we need to be open to other, less easily observable processes which may be at work and whose nonexistence is not conclusively demonstrated. The "gin and soda principle" (p. 66) is applicable here; that is, the most obvious correlation is not necessarily the real cause.

Races, Varieties, Breeds and Subspecies. As different populations of the same species adapt to various environments, they may become recognizably different in appearance and function. Depending upon the kind and amount of change, they are termed races, varieties, breeds or subspecies. Population changes of this kind and extent are well documented in nature and are also regularly produced by plant and animal breeders. This type of evolution may be called a fact.

Some creationists object to this type of genetic and ecologic change being called *evolution* because it does not fit their definition of evolution. John Moore acknowledges the scientific validity of these kinds of changes but says they are no more than genetic variations and should be labeled as such instead of confusing the student by calling them evolution.[18] Another creationist, Henry Morris, obviously does not define minor genetic change as evolution when he states "there is no real evidence for evolution in present genetic processes."[19] Evolutionists mistakenly believe that these creationists reject all forms of genetic variation when actually they are simply using different terminology.

Origin of Species

If this differentiation of races, varieties, etc., were to continue through thousands of generations, the next logical step would be the formation of different species. This process, which is called speciation, would occur if each variety became so different that it could no longer successfully interbreed with other populations of the same species. Some evidence indicates that this is the way certain plant species originate. But this evidence is not as clear as the evidence for the formation of varieties.

Thomas Emmel admits:

The thorniest problem in evolutionary biology is that of

> speciation. . . . Biologists have extrapolated from these observations the conclusion that in geographic race formation we are observing the beginning of the speciation process. . . . The distinction is made by many between *macroevolution*, the appearance of new species in time as we observe them in the fossil record, and *microevolution*, the development of new races (but *not* species) from already existing species through the gradual accumulation of genetic differences between geographically isolated populations.[20]

There is also some difficulty in deciding just when two groups of one population become two new species.[21]

Fifty years of genetic research on plants and animals with mutations have "never transgressed the limits of species."[22] Genetic research has been able to produce a new species experimentally only in some cases of hybridization. One scientist explains:

> Apart from some cases occurring in polyploid plants, no new species has ever arisen within the sphere of observation, let alone, "macroevolutionary" changes. Selection theory is an extrapolation, the boldness of which is made acceptable by the impressiveness of its basic conception. With a less picturesque theory, one would doubtless hesitate to expand cosmically and universally a principle which is controlled experimentally only to a rather limited extent.[23]

The transformation of species A to species B and B to C is what Darwin and paleontologists usually mean by speciation. Thus, species can evolve through time and even become a so-called new species without necessarily producing a greater number of species. This is like an extensive remodeling of an old house, new and different but still within the basic limits of the old.

Some creationists consider the famous fossil horse series as evidence for microevolution, not macroevolution.[24] The changes are all within the horse "kind" and are similar to those found today in living horses. Admitting that this amount of change occurs within a kind does not require accepting the entire theory of macroevolution.[25]

Much of the evidence for the origin of new types, some of which are considered species, is "instantaneous" speciation resulting from hybridization between members of different species. The hybrid offspring are usually much less fertile, sometimes even sterile. A good example of this type of hybridization is the mule—the sterile offspring of a donkey and a horse. When hybridization is followed by a doubling of the chromosome number (polyploid), however, fertility is often improved enough that the organism (usually a plant) is able to reproduce successfully in nature and retain its unique identity as a new species.

Most of the "new" characteristics in the hybrid or new species were already present in the parents. Brand new characteristics rarely appear. Thus, it is difficult to support traditional evolutionary theory using examples of this form of speciation (hybridization). A further complication in supporting the production of several new species from an old one concerns direction of change.

A comparison of hybridization with the type of speciation required for macroevolutionary theory is shown in Figure 6. The origin of "new" species as commonly demonstrated by experimental evidence (microevolution) is contrasted with the origin of species as hypothesized in macroevolutionary theory. Most of the characteristics of the new species in A are just intermediate expressions of characteristics already present in the parents. The situation shown in B can only be inferred as an interpretation of past events.

The formation of all past and presently existing species by macroevolution requires the situation shown in B, which has not yet been demonstrated experimentally. However, current examples of speciation by microevolution (A) can

Figure 6. Two Types of Speciation

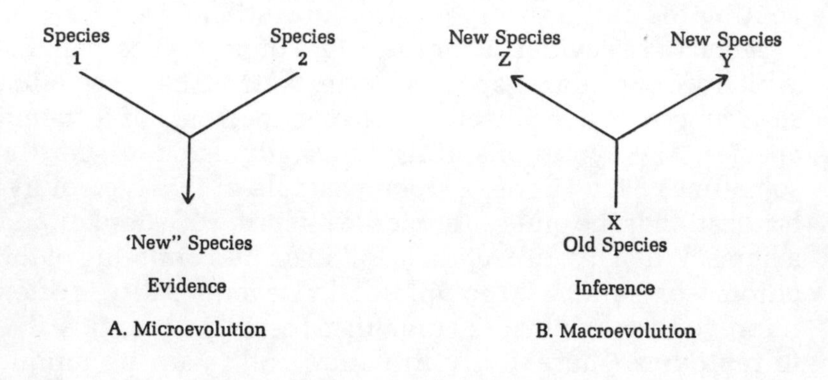

be demonstrated easily. It is estimated that over one-third of modern flowering plant species (and a few animals) are polyploid in origin.[26] In microevolution the old species (1 and 2) often persist as contemporaries of the new species which exists in intermediate habitats.

The Theoretical Side: Macroevolution

5

Despite their obvious differences, creationists and evolutionists do agree on many things. For example, they both agree that (1) the universe, life and basic forms came into existence in the past; (2) these origins were one-time events, not repeatable by us, nor were they observed by any human witness; and (3) we do not have and probably never will obtain direct knowledge of how these origins occurred, regardless of how many ways we demonstrate that they could have occurred.

To posit macroevolution, scientists have projected beyond microevolution and other current knowledge verified by repeated observation and experimentation. These projections, whether called ideas, speculations, extrapolations, educated guesses, hypotheses or theories, are a natural next step from our existing data. There is no problem in considering these projections for what they are, as long as they do not sneak into the false clothing of a fact."[1]

Origin of the Universe

The evidence of an expanding universe leads most cosmologists to believe the Big Bang theory. This theory states that the universe originated several billion years ago from the explosion of a small sphere of enormous density and temperature. During the initial rapid expansion, all of the present elements were formed and later contracted from the vast, rotating cloud of "dust" particles to form the planets, including our solar system.

The Steady State theory proposes a universe infinite in age and extent in which matter is continually created (out of nothing) to fill up the space left by expansion. Most scientists object to the creation of matter from nothing, since it violates the law of conservation.[2]

In his book *Concepts of Contemporary Astronomy*, Paul Hodge says that of the many hundreds of theories advanced to explain the origin of our solar system, only about twenty can be considered separate theories. He also states that "there is no complete theory of the origin of the solar system."[3] Thus, it seems premature to take a firm stand on any one theory. And certainly it would be erroneous to make any statement of fact about how our universe and solar system came into existence.

It is interesting to note that most of the accepted evolutionary theories start with something. In fact, evolution must always start with something. To evolve, something must change into something else. Special creation, on the other hand, begins with no material substance, only God.

George Schweitzer discusses the relationship of God, science and cosmogony:

> No person can come to the problem of cosmogony without preconceived ideas. Since one of the basic beliefs of Christianity is its concept of God as Creator, the question of cosmogony is particularly controversial in our society. Attempts to avoid the idea of creation are found throughout

the scientific literature, indicating the presence of much prejudice. Many writers are not content to leave the idea alone, which would be the strict scientific attitude since science does not deal with ultimate origins, but instead they take particular pains to set forth and promote belief in possibilities for a universe without an origin.[4]

Origin of Life

The origin of life is an essential requirement of macroevolutionary theory. This theory assumes the development of a simple life form from which other forms evolve. How much does science really know about the origin of life? One text states that "probably more words have been written on the origin of life by scientists with fewer facts and direct evidence at hand than any other topic in biology."[5]

Darwin spoke of life being breathed into one or more forms by the Creator.[6] Most scientists today reject this and are attempting to synthesize life from inorganic chemicals in simulated prehistoric environments. Other scientists, however, claim this is not realistic since the Precambrian environment is not known and assumptions of its nature are often contradictory.[7] In addition, the various compounds necessary for the synthesis require different environments for their synthesis.

The creation of living organisms from nonliving materials is spontaneous generation, an event which has been tried unsuccessfully for centuries. But, what if we do create life in a test tube? What will it tell us? For one thing, it will have been done by a designer, an intelligent being, and not by chance. And it will be an interesting, possible mechanism. But it will not tell us how life actually did originate. Such a simulated synthesis would be only one explanation among several, including special creation.

Why should scientists believe so firmly in spontaneous generation as the basic foundation for the whole theory, when it has not been demonstrated even once? Karl von Naegeli is probably closest to the truth when he says that "to deny spontaneous generation is to proclaim a miracle."[8] This is one thing most scientists try to avoid. Emmel admits:

> Clearly, no biologist will ever know the complete story of the origin of life. . . . A number of biologists feel that these hypotheses are compatible with a belief in Supreme Deity; others hold to a purely mechanistic viewpoint. Neither point of view can be proved by physical evidence; thus research in this area is intellectually interesting and stimulating but without promise of a definitive answer.[9]

Remember this when you read the factual, cut-and-dried discussion of the origin of life which often appears in high-school and college freshman texts. Not all standard texts err in this manner. But many of them do. The previous quotes have been included specifically to show creationists that some evolutionists tell it as it is. This quote from a standard botany text says it very well:

> The evidence of those who would explain life's origin on the basis of the accidental combination of suitable chemical elements is no more tangible than that of those people who place their faith in Divine Creation as the explanation of the development of life. Obviously the latter have as much justification for their belief as do the former.[10]

Origins of Groups above the Species Level

Although the origin of species may be for some a logical extension of the origin of races, varieties and subspecies, what about the higher categories? Have any of their origins been observed or experimentally demonstrated? What kinds of evidence exist for their origins? These questions and their answers are central to the creation-evolution controversy. Many scientists have spoken to this issue.

As one geologist points out, two kinds of evidence are used. "Although the comparative study of living animals and plants may give very convincing circumstantial evidence, fossils provide the only historical, documentary evidence that life has evolved from simpler to more and more complex forms."[11] Barber points out that even fossil evidence is circumstantial:

> Evolutionists regard the fossil record as a final court of appeal in substantiation of their theory. Whatever difficulties there may be in determining the how and why of evolution from the study of living things, always the evidence of the fossils stands to confirm the fact that multitudinous changes have occurred through the ages. It is in this light that fossils are often referred to as "the documents of evolution." . . . In all honesty it must be conceded that fossil series, too, are circumstantial evidence which permits an evolutionary interpretation and many people feel even demands it—but cannot in and of itself close the issue.[12]

Let us first look at the amount of fossil evidence available and then discuss its interpretation.

Most scientists agree that the fossil record is imperfect for several reasons: (1) not all organisms have bones, wood or other hard parts which are easily fossilized; (2) many organisms were not in situations where quick burial and protection from decay could preserve them; and (3) most relevant fossils lie undiscovered in deep or remote deposits. Yet there are numerous fossils which have been found, studied and used to support macroevolutionary theory. Let us examine some of this support by referring to scientific accounts.

In his book on evolution above the species level in flowering plants, Ledyard Stebbins writes:

> Taken at face value . . . the fossil record of angiosperms can be more misleading than enlightening as a means of interpreting the major trends of their evolution. Given the present state of knowledge, or ignorance, all that the

> botanist can do is to make logical inferences from such data as can be obtained from studying both living plants and the available fossils, and hope that at some future time facts will become available that will enable us to reaffirm or to reject the hypotheses that have been constructed on the basis of these inferences.[13]

Coleman and Olive Goin add that fossils of land plants appear much earlier than the algae from which they are said to have evolved.[14]

Animal fossils seem to be even more controversial. G. G. Simpson, who was mentioned earlier as one of the most prominent names in evolution, has written several books on the subject. Even though much more evidence has been discovered since the publication of his first book, his comments on the fossil record remain consistent and firm. In *Tempo and Mode in Evolution,* published in 1944, he wrote:

> The facts are that many species and genera, indeed the majority, do appear suddenly in the record, differing sharply and in many ways from any earlier group, and that this appearance of discontinuity becomes more common the higher the level, until it is virtually universal as regards orders and all higher steps in the taxonomic hierarchy.[15]

In *The Meaning of Evolution,* published in 1949 and revised in 1967, he repeats his assertions:

> There is little logical order in time of appearance. The Arthropoda appear in the record as early as do undoubted Protozoa, although by general consensus the Protozoa are the most primitive phylum and the Arthropoda the most "advanced"—that is, structurally the most complicated. ... For the present this failure of the phyla to appear in the order which would be expected as "natural" on the basis of increasing complexity is merely stated as a fact.[16]

In his work entitled *The Major Features of Evolution,* he is even more specific concerning gaps between categories above the species level:

> In spite of these examples, it remains true, as every paleontologist knows, that *most* new species, genera, and families and that nearly all new categories above the level of families appear in the record suddenly and are not led up to by any known, gradual, completely continuous transitional sequences.[17]

Other scientists say much the same thing: that the major groups of organisms appear suddenly with little or no transition. Everett Olson, another well-known evolutionist, mentions the periodic catastrophies of the fossil record which are cited by some creationists:

> More important, however, are the data revealed by the fossil record. There are great spatial and temporal gaps, sudden appearances of new major groups, equally sudden appearances of old, including very rapid extinctions of groups that had flourished for long periods of time. There were mass extinctions marked by equally simultaneous death of several apparently little associated groups of organisms. At the time the record first is seen with any real clarity, the differentiation of phyla is virtually complete. As far as major groups are concerned, we see little clear evidence of time succession in differentiation with the simpler first and the more complex later.[18]

Daniel Axelrod, a major contributor to fossil plant evolution, writes specifically about the absence of fossils in the Precambrian era, where transitions to these major groups should be found, if they exist.[19] Fossils have not been found in thick layers of suitable sediment beneath layers containing the earliest Cambrian fossils.

Some say that extensive geologic upheavals, pressures and heat have destroyed most or all of the Precambrian fossils. Most of the ancient rocks on earth, however, have not been subjected to metamorphism so severe that it would destroy fossils.[20] Several studies show that carbonaceous matter (a component of fossils) is a common constituent of metamorphosed sedimentary rocks. This matter apparently is not completely lost through normal metamorphic processes.

Thus far we have been discussing fossil evidence which applies mainly to plants and animals with bones. What about those simpler and presumably older organisms without bones? Bonner reveals that

> The particular truth is simply that we have no reliable evidence as to the evolutionary sequence of invertebrate phyla. We do not know what group arose from which other group or whether, for instance, the transition from Protozoa occurred once, twice, or many times. Most of us make the tacit assumption that the origin of life, and the origin of Protozoa themselves are unique events, but can we be sure? The evidence from fossils for these primitive groups has thus far been of no help.[21]

There are many fossils of some types but many wide gaps continue to exist. Other scientists echo the statements quoted above. Even more problems arise with regard to interpretation. What do these fossils mean? Do they support macroevolution? Or do they support creation? Can they support both viewpoints?

Interpretation of Fossil Evidence

Various specialists in the field of evolution have spoken on how the fossil record is interpreted. Barber points out that fossils do not necessarily support just one hypothesis. They may be rationally interpreted in several ways:

> To a large extent the basic philosophy of an individual will enter into his consideration of fossils. As one evolutionary professor of paleontology remarked, "You can 'prove' almost anything you want to from fossils. . . ."
>
> Fossils do not prove evolution. Neither do they disprove it. They strongly suggest that a considerable amount of descent with modification has transpired. They also exhibit a lack of transitional forms which may or may not

be significant of limits of relationships.[22]

This gives us some insight into the reasons for the occasional misinterpretation of fossils. Perhaps the fossil peccary tooth was first interpreted as belonging to a prehuman form because of an overriding interest in obtaining evidence of human ancestry. Or perhaps it was just a human error in determining the identity of an unknown tooth.

Dwight Davis thinks that "the facts of paleontology conform equally well with other interpretations that have been discredited by neobiological work, e.g., divine creation, . . . and paleontology by itself can neither prove nor refute such ideas."[23]

There seems to be two fundamentally different ways of interpreting the fossil evidence. A *monophyletic* interpretation asserts that all the major groups originated from one source. A *polyphyletic* interpretation, on the other hand, proposes that several sources are required. (See Figure 7.) Those who espouse a monophyletic interpretation are not deterred by the large gaps in the fossil record. They believe the record contains adequate evidence of transitions between the major groups to be able to support the assumption that they had a common source. Many evolutionists choose a monophyletic interpretation based on the evidence of biochemistry. That is, different species within the same order have a similar biochemical composition. For example, the biochemical composition of the hemoglobin molecule in humans, gorillas, horses and pigs is quite similar. This similarity can be explained by a common origin (evolution) or a common design (creation). Other explanations of this phenomenon are also possible.

Those who do not believe that there is enough evidence of transition from one species to another opt for a polyphyletic interpretation. Daniel Axelrod says that the major categories of vascular plants "may well represent unrelated groups that developed independently from different algal sources."[24]

John Moore believes that genetics also supports a polyphyletic interpretation. He asserts that specialists in the study of evolutionary development "are duty bound to research and write in multiple, hypothetical fashion of the contrast between a monophyletic interpretation of facts and a polyphyletic interpretation of the same facts. Furthermore, science teachers and professors, who use the results of research specialists, should be duty bound in *academic freedom and responsibility* to present both monophyletic and polyphyletic interpretations."[25] This exhortation has largely gone unheeded.

The Influence of Fossil Dating

The amount of time permitted for the origin of life and the development of major groups is a most important variable. Most evolutionists believe that given millions of years, even the most improbable events are bound to happen sooner or later. Extremely long periods of time are essential to macroevolutionary theory but not to a theory of creation, which is interpreted by many creationists as having occurred as recently as six to ten thousand years ago. Both creationists and evolutionists have their own good reasons and evidence for their respective time periods.

Evolutionists use several methods for dating fossils, all based on untestable assumptions. The most widely used method is based on the decay rate of several kinds of radioactive materials, such as the uranium-lead or potassium-argon "atomic clocks." An age is estimated by measuring the amount of uranium remaining in a rock sample, comparing it with the amount of lead formed and multiplying this by the decay rate.

This method of dating fossils assumes several things, including: (1) only the radioactive material, and not any intermediate or final decay products (for example, lead) were initially present; (2) no intermediate or final product was added or lost since its initial formation (although some intermediate products are gases); and (3) the rate of radioactive decay has not varied since the beginning of time. Actually, what the scientists determine is only the amount of initial and decay products remaining in the sample. Determining the age requires calculations based on the above assumptions. If these assumptions are true, the calculated age is realistic. But until these assumptions can be supported, no dates based upon them can be known with certainty. There is some question as to just how far the present can be extended into the past, especially when billions of years are involved.

Some creationists use biblical genealogy to calculate their time scale. They are doubtful of the assumptions used in radioactive dating and the extremely long dates resulting from them. They are also aware of the fact that some scientists report that coal and petroleum can be formed in a matter of hours, not millions of years, if plant materials are subjected to heat and pressure.[26] Creationists also refer to articles suggesting that atomic clocks can be reset by neutrino fluxes[27] and that cosmic dust influx and the age of comets do not indicate an old earth.[28]

Therefore, although the earth appears to be and could be very old, the problematic nature of several assumptions gives reason to doubt this theory. Your conclusions will rest on the set of assumptions in which you have the most faith. Or you can leave the question open and decide when more evidence is available.

Origin of the Human Race

Much has been said about the origin of man. Most of these statements show more faith than evidence and sometimes more emotion that thought. What has been said about the origin of man that is not based on faith? What factual statements can be made? One scientist, Bernard Campbell, stated (p. 65) that there were no directly pertinent facts, only inferences and hypothetical interpretations of indirect evidence. Another writer expresses the improbability of man's evolution by saying that "considering the enormous number of heritable traits that were established (and later changed) in adaptation to special circumstances in the long line of human ancestry on earth, man must be one of the most improbable productions of this universe."[29]

Faced with the choice of faith in highly improbable events or faith in God's Word, creationists choose the latter. Some creationists, aware of the lack of sufficient fossil evidence to connect man to animal ancestors, believe that man is not genetically related to other forms of animal life. Animals were made after their own kind and man was made in God's physical and spiritual image. Others believe that, based on John 4:24, people were created only in God's spiritual image.

The fact of man's possession of certain features common to all life is interpreted quite differently by creationists and evolutionists. Some evolutionists unfamiliar with the Bible assume that if God created, he would have created randomly, without order or system. Thus, any evidence of common features is believed to be evidence contrary to creation.

Creationists, on the other hand, emphasize that the presence of common features and design imply a designer, not chance events. It seems quite reasonable to say that God created organisms with similar features because they would inhabit the same earth. But he allowed for variations on a theme to take advantage of environmental changes. For example, the people who design our cars and houses use variations on a common theme to accommodate different functions.

Phylogenetic Trees

Phylogeny refers to the development of and relationships between major groups of organisms. The illustrations used to represent these developments are called phylogenetic trees because of their appearance. Your family tree is one type of phylogenetic tree.

Because information on the relationships between major groups of organisms is not always available, phylogenetic trees are constructed on a hypothetical basis. In fact, based on the evidence for the development of any one particular organism, several phylogenetic trees can be constructed. (See Figure 7, p. 113, for several possible phylogenetic trees based on the same evidence.)

In regard to human origins, Ian Tattersall and Niles Eldridge urge those who believe in human evolution to return to the basic data (fossils). Only at this level are there testable conclusions. The greatest amount of agreement is also found on this basic level. Tattersall and Eldridge state that "in moving to the more elaborate level of the phylogenetic tree, we are moving essentially into the realm of speculation, informed or otherwise."[30]

Each phylogenetic tree can, in turn, give rise to several scenarios. A scenario is similar to a phylogenetic tree, but it includes information about adaptation and environment. Because it gives more information but is based on the same amount of evidence, a scenario is even more hypothetical than a phylogenetic tree. It is, therefore, more subject to personal biases. But it is often the scenario which is taught to biology students. And, it is presented to students as fact, not theory. This conversion of theory into fact is objectionable to creationists and some evolutionists.

Louise Young points out the effect of human tendencies by saying that "mankind yearns for certainties and the non-scientist, looking at the miracles which science has achieved, is apt to attribute to scientific discoveries a finality which they do not possess. Sometimes this confusion is compounded by the enthusiastic scientist who presents his favorite theory or discovery in such terms that it sounds like an irrefutable fact."[31]

Phylogenetic trees should be considered for what they are—the best ideas available, not demonstrated facts. Herbert Ross speaks of phylogenetic trees as complex models.

> Now that we have brought the *complex model* into the picture, we must add an extra word of caution. A complex model is simply a compounding of simple models or hypotheses. . . . If we insist on admission of new evidence, there is always a chance of demonstrating that any of our premises are either faulty or incomplete. For this reason, all of our conclusions can at the most be hypotheses. This means that, although we seldom say it, we should always realize that the word "if" precedes all of our assertions.[32]

As we have stated previously, one of the questionable assertions is that microevolutionary processes are sufficient to produce macroevolutionary change. Most modern evolutionists extrapolate beyond the direct evidence of microevolution to a macroevolutionary theory of the formation of higher categories of organisms. Kerkut provides some perspective on this:

> Much of the evolution of the major groups of animals has to be taken on trust. There is a certain amount of circumstantial evidence but much of it can be argued either way. Where, then, can we find more definite evidence for evolution? Such evidence will be found in the study of living forms. It will be remembered that Darwin called his book *The Origin of Species* not *The Origin of Phyla* and it is in the origin and study of the species that we find the most definite evidence for the evolution and changing of form.[33]

The question of whether or not microevolutionary processes through time are sufficient to account for the possibility of macroevolution cannot be answered directly. The sparse fossil evidence and inability to test hypotheses concerning past trends make it impossible to develop conclusive phylogenetic trees. Yet phylogenies can project general, tentative trends.

J. T. Bonner warns that phylogenies should not be given more emphasis than they are due. Their tentative nature must always be kept in mind.

> The great phylogenetic schemes, no matter how delicious and tempting, must wait. They must wait because our present evidence is inadequate to decide between schemes, and working hypotheses lose their glitter if there does not seem to be any possible means of critically testing them. But, even more important, they must wait because there are more fundamental problems to solve first.[34]

Bonner goes on to say that with regard to phylogeny, textbooks tend to be a "festering mess of unsupported assertions."[35] Scientific creationists are therefore justified in their fight for science texts which either simply state the evidence without presenting a theory or present several competing theories.

An evolutionist who was earlier cited because of his emphatic insistence on the factualness of evolution admits that a completely satisfactory mechanism to explain macroevolution has not been found. In his book *Evolution*, J. M. Savage states:

> Is the grand pattern of evolution only the result of simple population change? To many paleontologists, and to those biologists interested in major evolutionary shifts, the question remains open. No satisfactory mechanism or mechanisms have been proposed that might explain these phenomena, but the characteristic modes, patterns, and pathways of evolution at this level all suggest that other factors besides those operating at the population level must contribute to adaptive radiation and to the origin of new biological systems.[36]

Savage also asserts that there are fewer than two hundred major biological organization plans developed in the last three billion years, most of which still exist.[37] This number could just as easily refer to the number of created kinds.

Why Not Creation?

Much of the evidence about origins can be interpreted to support either macroevolution or creation. Which one should be chosen and why? E. J. H. Corner offers this explanation of the evidence:

> Much evidence can be adduced in favour of the theory of evolution—from biology, biogeography, and paleontology, but I think that, to the unprejudiced, the fossil record of plants is in favour of special creation. If, however, another explanation could be found for this hierarchy of classification, it would be the knell; can you imagine how an orchid, a duckweed, and a palm have come from the same ancestry? And have we any evidence for this assumption? The evolutionist must be prepared with an answer.[38]

Recent research still has not produced the evidence called for by Corner. The evolutionist crosses these gaps by faith in evolution in the same way that a creationist crosses the gaps by faith in God. It is not a matter of whether or not one has faith, because either choice requires it. It should be understood that the object of faith is one of the biggest differences between evolutionists and creationists. Figure 7 shows three different interpretations of the evidence of the origins of major groups of organisms. Some evolutionists interpret the fossil evidence as monophyletic with a common origin (A). Others interpret the evidence as suggesting many different origins (polyphyletic) with subsequent differentiation (B). Creationists interpret the beginning of the fossil record as the time of creation of the basic kinds of organisms which have since diversified (C). Dotted lines represent large gaps between major groups of organisms. These gaps are bridged by faith. Many texts err by indicating that there exists an abundant, continuous fossil record (solid lines) and the Precambrian era where, in fact, fossils are few.

Figure 7. Three Interpretations of Evidence of Origins

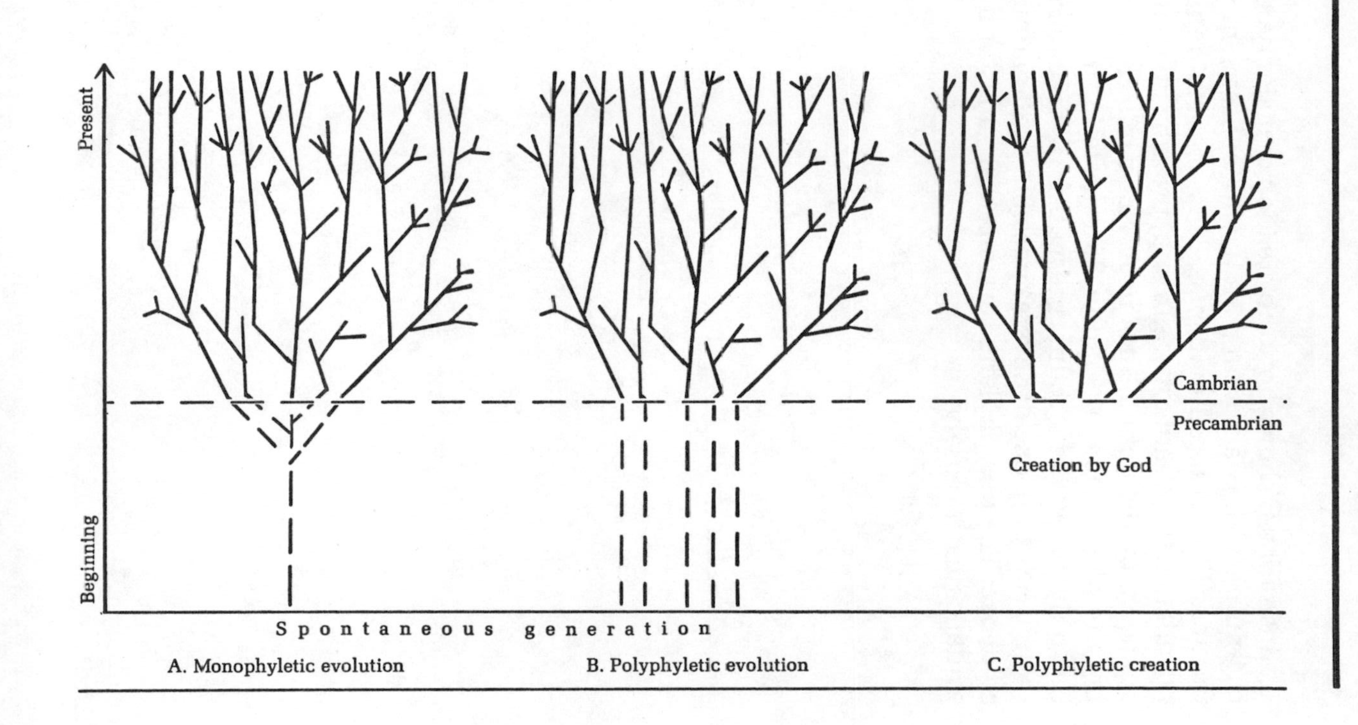

Illustration C shows how the creation of major types at the beginning of the Cambrian period, with subsequent diversification, fits the fossil evidence. This does not require the postulation of a common origin supported by missing fossils as shown by the dotted lines in A. Nor would it require belief in spontaneous generation. Interpretation A is usually the only theory of origins presented in textbooks, even though it is only one of several possible explanations of the evidence. The polyphyletic versions of B and C fit the fossil evidence most closely.

Creation

6

The origins of matter, life and people are included in Genesis and many other books of the Bible. Yet the Bible is not a science book. It tells the "why" and "who" of creation but little about the "how," which is science's domain. The Bible is quite accurate and specific about some scientific matters, such as sanitary practices and the most appropriate day for circumcision, but its main message tells of our relationship to God and to others.

Creationists venture into the "how" of creation by developing several different interpretations. What the Bible leaves out, men and women are eager to supply. Six versions of creation theories will be briefly discussed here. Others are mentioned in *A Christian View of Origins* by Donald England[1] and *100 Questions about God* by J. Edwin Orr.[2]

Each version is scientifically or theologically possible, although each one also has certain weaknesses. For example, one version may be unable to provide suitable answers to important questions while another version has apparent inconsistencies. Each version's supporters handle most of these deficiencies by relying on a sovereign God who is capable of solving them. We are more concerned here, however, with what God actually said or did than what he could or should have done to make some interpretation more acceptable.

Biblical Interpretation

Some of the different versions of creation were developed because of differences of opinion on whether the biblical account of creation is historical fact or religious symbol. Some consider the Genesis creation account to be "paragraph pictures" (that is, pictures in words).[3] Others quote Jesus and Paul for evidence that the account is historical.[4]

Those who accept the creation account as historical are often thought of as holding a literal interpretation. These interpretations, however, are unique, rather than literal because some of their differences come more from their additions to Scripture than from actual scriptural text. For example, some people who hold to a literal interpretation date creation at 4004 B.C. But the phrase "in the beginning" does not specify a date. So a truly literal interpretation would leave the date indefinite. Any date is derived by deductions, assumptions and inference added to the text.

Other differences in interpretation arise from the various meanings given to certain words such as *created, dust* and *day*. The interpretation of these words is largely a matter of judgment which uses context, usage and reason. It makes a big difference to the supporters of some theories whether something was created from nothing (Hebrew *bara*), made from pre-existing matter (Hebrew *asah*) or brought forth by the earth. The way these and other terms are interpreted will determine what particular theory of creation will be adopted.

There are some general guidelines which many scholars agree will be helpful in interpreting difficult biblical texts.

Guidelines for Understanding the Bible

The Hebrew scholars producing the new translations of the Bible must decide whether to start Genesis 1:1 with "In the beginning God created the heavens and the earth" (RSV), or "When God began to create the heaven and earth" (Anchor Bible). The problems you will face are in deciding what your English translation means. Christians are told to "do your best to present yourself to God as one approved, who has no need to be ashamed, rightly handling the word of truth" (2 Tim. 2:15). This is not an easy task. It should be approached with humility and begun with prayer. The following guidelines may aid your study.

1. *Recognize the influence of your biases, presuppositions, ignorance and background.* These can both prevent you from seeing what is in the Bible as well as cause you to believe things that are *not* in the Bible. The date of 4004 B.C. is an example of a belief which is not based strictly on the biblical text. Remember that you are trying to find out what the Bible says about a certain topic, not reinforce your notion of what you want it to say.

2. *Study the passage in context.* Each passage should be studied as part of a whole—a whole paragraph, chapter, book. Many heretical positions have been supported by Bible verses which were used out of context. Since the whole Bible is a single revelation from God, individual passages should be interpreted in the light of the entire message.

Knowing the history and culture of the particular time and place can also be a great help. A good Bible handbook is useful. Remember, you are trying to learn the truth, not just find a few scattered pellets of biblical buckshot to defend your beliefs.

3. *Be willing to change your mind.* You may discover new evidence, that is, passages that you did not know were in the Bible which clearly indicate that your first idea had little or no biblical basis. The evidence for your interpretation may

not be strong enough to justify a firm stand. Or perhaps the point in question is revealed to be so minor that it is not worth arguing about. Try to be objective and use common sense in these matters.

4. *Admit your lack of knowledge.* Be willing to admit that you do not know exactly what a passage means. Some passages may be very difficult to interpret because they allow so many interpretations. Or perhaps you do not have the background necessary to understand a verse. The mature and scholarly action in these cases is to freely admit the difficulty and investigate it further, with as much fairness and as little bias as possible. Eventually, you may decide that one interpretation is more valid than another. And if you make up your own mind, you will know the reasons behind your decision.

5. *Examine all the relevant Scripture.* You should not be content with examining only one or two passages of Scripture on a particular subject. You should use a concordance or other Bible study aid to help you to find all of the relevant passages. If you limit your study to only one or two passages, you may be getting only one aspect of what the Bible has to say. Other verses give you a much wider perspective.

6. *Be aware of the various uses of language.* Biblical scholars have pointed out that we should read the Bible in the same way that we read other books:

> We distinguish between lyric poetry and legal briefs, between newspaper accounts of current events and epic poems. We distinguish between the style of historical narratives and sermons, between realistic graphic description and hyperbole. Failure to make these distinctions when dealing with the Bible can lead to a host of problems with interpretation. Literary analysis is crucial to accurate interpretation.[5]

We cannot discuss at length the various types of language used in the Bible. It would be best for you to spend some time on your own studying biblical interpretation.[6] But for now you must at least recognize that many of the disagreements between Christians about origins stem from the various ways of interpreting Genesis.

Time and Biblical Creation

There are two questions about time which are of primary interest here. These are the questions of the duration and the date of creation. They will probably never be agreed upon because the biblical text does not give a date for creation. David Riegle believes that since the date of creation is not given, God must have considered it of little importance and therefore any date given will be mere speculation.[7] Yet Bishop Ussher calculated the date as 4004 B.C. And other medieval scholars, it is reported, apparently calculated the date as 9 A.M. on October 23, 4004 B.C.[8] How did they do it?

These dates are based on the assumption that the genealogies (lines of descent) and ages of the patriarchs given in the Bible are accurate records of elapsed time. Many people, however, have misunderstood the purpose and structure of genealogies. A biblical genealogy is a study of important members of a family or line of descent and does not necessarily include every member of that line. Francis Schaeffer points out that Scripture makes it clear that the genealogies are not intended as chronologies.[9] For example, Matthew 1:8 omits the names of Uzziah's father, grandfather and great grandfather which are given in 1 Chronicles 3:11-12.

Furthermore, Jesus is often referred to as the "son of David," and the Israelites as "sons of Abraham." Clearly sonship is used here in a general sense. To assume continuous lineage is to invite error. Some non-Christians even use the examples mentioned above to prove that the Bible cannot be trusted. But the problems do not arise from errors in the Bible. They arise from not recognizing the purpose of biblical genealogies. These genealogies are not included as exhaustive lists, but merely as indications of family relationships.

The relation of genealogy to time is summed up well by Paul Zimmerman:

> The Bible does not anywhere make an explicit statement in which the age of the earth is given. It tells us how long the Children of Israel were in Egypt, the length of time from the Exodus to the building of Solomon's temple, the duration of the Babylonian captivity, etc. But nowhere is there a statement of how many years it was from the creation to the time of Abraham or any other date that be correlated with secular history. It is important to remember this point. Any estimate of the age of the earth based on the Bible rests on deductions drawn from information contained in Holy Scriptures.[10]

Many who believe in a so-called young earth of about six to ten thousand years will encounter problems. These problems include difficulties in interpretation as well as difficulties in accounting for geologic evidence which indicates that the earth is ancient. These problems cannot be resolved because of the lack of agreement in biblical interpretation and the lack of acceptance by creationists of the many assumptions required in geologic dating. As stated earlier, these assumptions cannot be proven. But extremely long periods of time are necessary to allow the possibility of highly improbable events suggested by evolutionists for the origin of life.

The six versions of creation which we will discuss here are grouped according to their interpretations of time.

Creation in Six Twenty-four-hour Days

Ruination-Reconstruction or Gap Theory. This is a widely held version of creation. The original creation of the first perfect earth is described in Genesis 1:1. This is followed by a vast time gap in which the organisms now known as fossils were living. Genesis 1:2 describes the death of all life caused by Satan, who brought destruction to the earth after being cast out of heaven. The six twenty-four-hour days of Genesis 1:3-31 would then actually be days of re-creation in which the earth was made perfect again with new animals and plants. Man, the second ruler of the earth, was tempted by Satan, fell, was judged, and the earth was cursed again.

This interpretation permits full acceptance of the fossil record, the geologic time scale of millions or billions of years, and also the six twenty-four-hour days of creation about six thousand years ago. Genesis can be considered both historical and factual, thus providing a strong base for the salvation message.

This view is not, however, without problems. First of all, there are no supporting verses from the remainder of the Bible; that is, the rest of Scripture does not indicate that Satan destroyed the original creation.

Second, Exodus 20:11 says, "for in six days the LORD made heaven and earth, the sea, and all that is in them, and rested the seventh day; therefore the LORD blessed the sabbath day and hallowed it." It neither states nor implies re-creation to most people.

Third, if re-created, why are only a few of our present forms of life apparently identical to the forms of the previous creation as seen in the fossil record?

Finally, if we accept the fossil record, how do we explain the presence of humanlike remains? Did people exist in the first creation as well as the second? Were the "men" found in the fossil record just animals without a soul?

Flood Theory. A second widely known version of creation is generally referred to as the flood theory or flood geology. This is the version strongly supported by the Creation Research Society and many others who are actively involved in the science textbook controversy. Consequently, it is included in numerous books, such as: *The Genesis Flood; Fossils, Flood and Fire; Scientific Creationism;*

and *The Great Brain Robbery.*[11] This version is the only one which strongly contradicts the geologic time scale. And its supporters are probably the most vocal and dogmatic.

Briefly, the flood theory states that the earth and its life were created in six twenty-four-hour days about six to ten thousand years ago. Much of the fossil record consists of remains of preflood animals which lived in quite a different world. Due to the catastrophic actions of the flood, the simpler, less motile organisms were buried first (and are therefore found in the lower layers of sedimentary rocks). The more complex, motile organisms reached higher ground and were buried in the upper rock layers. Present forms of animals are descendants of those saved from the worldwide flood in Noah's ark.

This account appears on the surface to adequately handle both the six twenty-four-hour days of Genesis and the fossil record. But it also raises numerous problems.

The first problem is that plants show the same general trend of less complex forms in the lower rock layers and more complex ones in the upper layers. Yet none of the plants are motile, and therefore they cannot reach higher ground. Although the creationists have offered explanations for this—the plants floated to the top or were stratified in ecological zones—none of these explanations seem adequate.

Also, the Bible specifically mentions destruction of all that moved on the earth and on the dry land; however, nothing was mentioned to explain the destruction of the sea creatures, which would be expected to survive quite well during the flood.

Finally, how did the plants survive such a destructive flood and become re-established so quickly that the dove could bring back an olive leaf? A worldwide flood which buried both plants and animals under sediment sometimes thousands of feet deep makes this highly improbable. Although a few plants can survive salty waters that long and a few could colonize raw sediment, most cannot. The very principle of plant succession shows that most plants must wait many years for the development of soil and appropriate moisture and shade conditions before becoming established.

Pictorial or Revelatory Day Theory. This version proposes that the six days of creation are days of revelation in which the creative events, which actually happened over very long periods of time, were successively revealed to man, each in a twenty-four-hour period. The fact that each revelation took place during a twenty-four-hour day is said to accomodate the references to evening and morning in the biblical text.

Critics point out, however, that elsewhere in the Bible revelations are mentioned as such. In most cases, no doubt is left in the minds of readers as to whether a passage is revelatory or historical in nature. Further detail on this version, which is similar to the day-age theory, can be found in the book, *Creation Revealed in Six Days.*[12]

Appearance of Age Theory. The appearance of age theory is important to understand because some evolutionists use it as a representative creationist view. Martin Gardner included it (under the name of *divine deception*) along with the flood theory, in his book *Fads and Fallacies in the Name of Science.*[13]

This view was proposed by Philip Gosse, a zoologist who wrote a book called *Omphalos* (the Greek word for navel). Gosse chose this unusual title for his book because his theory was loosely based on the assumption that God created Adam with a navel.

This theory states that God formed the earth several thousand years ago, but gave it the appearance of being much older. The argument rests on the belief that God created the earth and its inhabitants in a mature, fully functional form. Thus, when God created trees he probably created them with rings even though they had not grown. Likewise, even though Adam and Eve were not born of biological parents, they presumably had navels, evidence of a birth which never occurred.

This theory is not, however, without severe problems. The main problem is, of course, that this theory implicates God as a divine deceiver. This concept of God certainly does not mesh with Scripture which describes him as the giver of truth.

Furthermore, if God has deceived us by making the earth appear much older than it is, how can we be sure that he is not deceiving us about other things? Perhaps he really created the earth a few seconds ago. How can we be sure that other people exist? Perhaps the external world and other people are all simply part of a dream in which we are involved. This view ultimately leads to total skepticism.

Creation in Geologic Time

Theistic Evolution. This version is the most popular one among those who accept both Christianity and evolution. The phrases in Genesis, *let the earth bring forth* and *let there be,* suggest that the various life forms developed from the basic materials of the earth. Theistic evolutionists believe that God first devised the natural laws which would govern the universe. He then created matter, the heavens and the earth. Life developed through a gradual, evolutionary process.

Like the others, this view is also open to criticism. First, how could God declare such a slow, wasteful, inefficient, cruel and mistake-ridden process to be good? How could a God of love use natural selection, in which the weak lose out, to create his perfect world?

Second, Genesis 2:7 states that man was made from the dust of the ground, not evolved from creatures which had developed earlier. The origin of Eve presents an even greater contrast between evolutionary theory and the biblical text.

Third, Adam is a necessary entity for the biblical message of salvation. A comparison was made between Adam and Jesus Christ in 1 Corinthians 15:22-23, 45. Since the evolutionary unit is a population, not an individual, how do we handle the singularity of this passage? If God chose one individual (Adam) from a population of evolved manlike beings and gave him a living soul, what happened to the rest?

Finally, macroevolution still lacks sufficient evidence to consider it factual. This view is questionable because it relies on evolution as the mechanism for the origin of life.[14]

Day-Age and Progressive Creation Theories. These two versions will be considered together because of their similarities. Both involve periodic intervention by God, followed by long periods of geologic time during which development occurred. Each also has its minor variants in name and form.

The day-age theory was developed in the early 1800s by Baron Cuvier, the famous French naturalist who also established comparative anatomy and vertebrate paleontology as scientific disciplines. He proposed a series of separate creations through geologic time, each followed by a catastrophe (for example, the flood) which buried the earlier forms of life. Thus, there would be six periods (days or ages) of indefinite length during which life evolved from created kinds with only periodic intervention by God. The Hebrew word *yom* is translated to mean age, a usage employed elsewhere in the Bible.

The progressive creation theory is essentially the same as theistic evolution except that it postulates six days on which God intervened in the natural processes in order to initiate new forms of life.

Many critics of these theories believe that the phrase, "there was evening and there was morning, one day," should be applied *only* to a twenty-four-hour solar day, not to geologic ages. This theory also shares the criticisms of theistic evolution to which it is closely related.

Additional discussions of these theories may be found in Stoner's *Science Speaks*, Bernard Ramm's *The Christian View of Science and Scripture*, R. L. Mixter's *Evolution and Christian Thought Today*, and Henry Morris's *Scientific Creationism.*[15]

Your Approach to Controversy

7

How should you solve the creation-evolution and other Bible-science controversies? The solution is to make a careful, comprehensive study of the problem according to the guidelines given in this book. This will help you to integrate your knowledge with your faith to produce a satisfying answer. In matters which are open to discussion (such as the meaning of time in the biblical text or the importance of gaps in the geologic evidence), it is my hope that you will grant others the same freedom and responsibility that you would want others to give you. It is better to "agree to disagree" than to errect barriers to communication. You may not be able to accept other people's interpretations, but you should continue to treat them with love and respect.[1]

You have now been exposed to a wide spectrum of thought on the origin of life, including direct statements from many specialists in the field. This should better equip you to make a decision on origins and to know why your choice was made. Your choice will be based on some form of reasoned faith. All of the choices are based on faith, whether in God and his creation or in the primacy of natural laws.

Relying on faith does not mean that your mind is useless. Both creationists and evolutionists use faith to go beyond the limits of physical evidence. This chapter will briefly summarize some of the main points discussed earlier.

The Renewed Controversy

Chapter one described the textbook issue in California which reawakened the creation-evolution controversy. Unlike the famous Scopes trial in 1925 which challenged the suppression of evolutionary theory, this conflict flared because creation was not being taught as an alternative to evolution. Although presenting alternatives is good scientific practice, science teachers and some scientists reacted strongly against the new requirement to teach alternatives to evolution.

The creation-evolution controversy is good for science because it challenges scientists to produce evidence to support their claims. Had Kerkut's and Bonner's calls to action been taken seriously, this controversy might not have arisen. Unfortunately, too many scientists and science writers reacted against the creationists instead of responding to their challenges. This was one of the clearest signs that macroevolution had a religious aspect. The creationists had boldly asked where the "emperor's clothes" were and startled a lot of complacent people.

Science, Evolution and Origins

Our study of science has revealed that it is powerful, but limited. For example, science cannot disprove creation. In this sense the various theories of creation are much like macroevolutionary theory. They are all assertions which go beyond the experimentally verifiable level and become untestable theories.

Science is based on both faith and logic, but it is not limited to either. Deductive reasoning from inductively derived hypotheses can at most be highly probable. The inability to test the assumptions underlying much scientific reasoning prevents any conclusion from achieving the certainty implied by its logical validity. Faith is required, whether the theory is creation or macroevolution.

The tendency is strong for science writers to make assertions whose certainty exceeds the evidence. And this tendency is often seen in science textbooks. Positive results may support a hypothesis, but never prove it. We should constantly remind ourselves not to place excessive trust in even widely accepted theories.

All scientific work that is done according to correct scientific methods can be accepted without fear of contradicting the biblical text. The problem lies in determining which aspects of science are reliable and which are not. Macroevolution is a very theoretical part of science. That is, the indirect, circumstantial evidence which supports macroevolution can be interpreted in many different ways. Microevolution is much less theoretical and can be accepted more readily.

No scientific facts compel you to accept either creation or macroevolution. These different interpretations of the same evidence are matters of choice, regardless of what some textbook authors say.

Figure 8 illustrates the relationship between the common evidence (microevolution) and the interpretations based on it. The choice of any particular version of creation or evolution depends largely upon the basic assumptions, interpretations and faith added to the evidence. Theistic evolution is shown as both a form of creation and a version of evolution. This is because some theistic evolutionists accept spontaneous generation while others do not.

Figure 8. The Relationships between Versions of Creation and Evolution

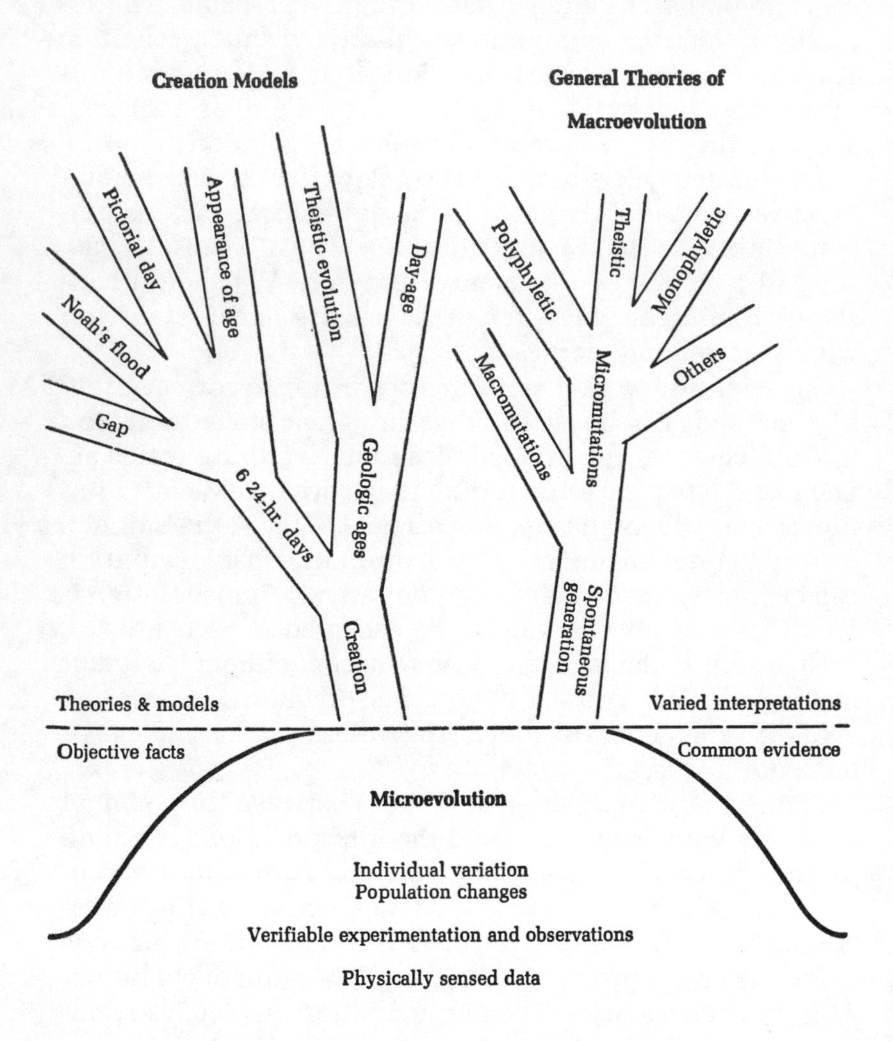

The Scientific Evidence

There is scientific evidence for three types of evolutionary change:

Individual Variation. This refers to change within populations and among offspring of the same parents. This form of variation can be demonstrated by laboratory experiments on animals. It also occurs in the natural world. There is no conflict between Scripture and the concept of individual variation. This process is accepted as factual by both creationists and evolutionists.

Microevolution. This refers primarily to change between populations (for example, the development of breeds of milk cows). This process, because it can be repeated under controlled conditions, is also accepted as a fact by both creationists and evolutionists. Some creationists, however, prefer not to call this evolution. There appears to be no conflict between the Bible and the concept of microevolution.

Macroevolution. This term refers to the theory that life began by spontaneous generation and the higher orders of organisms originated by evolution. These groups (such as genus, family, order) may have evolved from several sources (polyphyletic version of the theory) or they may all have originated from a single source (monophyletic version). This theory is usually interpreted as contradicting Scripture because it assumes that life originally came into being through spontaneous generation. The Bible, on the other hand, makes it clear that God was responsible for creating life. Furthermore, this theory is not well supported, except by evidence which is largely circumstantial and indirect. Because of this, some evolutionists, as well as creationists, do not accept this theory as a statement of fact.

Just as macroevolutionary theory cannot be demonstrated with absolute certainty, so the Bible does not try to *prove* creation. Two Christian biochemists report that the mechanisms used in the creation of life are an open question because the Bible gives little specific information. They add:

> Scientific investigations must always be mechanistic in their outlook because it is only in this narrowed frame of reference that the scientific method can operate. It is probably true that many nonbelievers have welcomed mechanistic interpretations of phenomena, and especially of the origin of life, feeling that such interpretations would make belief in God unnecessary. Such an attitude is indicative of a basic misunderstanding of the Christian idea of God and the way by which we come to know Him. Belief in God is never forced upon us, no matter what our level of understanding of natural processes. God cannot be found by scientific knowledge any more than He can by scientific ignorance.[2]

Scripture, Creation and Origins

People who believe in God, the supernatural and the Bible approach the evidence with assumptions different from nonbelievers. Consequently, their interpretations are also different. Differences among creationists arise mostly because of varying interpretations of the biblical text. Beyond establishing that God created in the beginning, many of the details about how this happened can only be inferred from the Bible. These details are not given and probably should not be the subject of serious speculation. Although the question of the origin of life is very interesting, the high improbability of finding a firm answer should encourage people to direct their study to more fruitful topics.

We have discussed some of the varying presuppositions held by those who are involved in the conflict over origins. Some people deny that any supernatural mechanism is needed to explain the origin of life. They believe that this can be explained on the basis of purely natural mechanisms. But they recognize that the means by which life began cannot be absolutely proven. Nevertheless, they have faith that macroevolution is correct in principle. Because of this faith in natural laws and a lack of faith in God, those who espouse naturalism are limited to mechanistic explanations for the origin of life.

Christian theists, on the other hand, have a faith in God which is greater than their faith in natural laws or the methods of science. This faith will often lead scientists who are Christians to conclusions which differ from the conclusions of their non-Christian colleagues.

Yet even within both the camps of naturalism and theism, there are differences. Some scientists who believe in microevolution do not accept some versions of macroevolution. They hold presuppositions different from their colleagues. They believe in evolution as the mechanism by which life originated, but theorize that evolution began with many life forms (polyphyletic interpretation of the evidence for evolution). Others believe that the evidence supports the theory that all life evolved from a single form (monophyletic interpretation).

Christians also differ among themselves. They all look at the same evidence—the Bible and scientific data—and come to different conclusions. That is, they interpret the evidence in different ways. Some Christians assert that the Bible indicates that life was created in six twenty-four-hour days. Others say that Genesis must be interpreted in a general sense and insist that God used evolution as his means to create the world. It was a miraculous process initiated and guided by God.

The creation-evolution controversy is indeed complex. It is not divided evenly between Christian creationists on the one hand and non-Christian evolutionists on the other. There are Christians who believe in macroevolution and non-Christians who think there is not enough evidence to support this theory. This book should make it clear that we cannot reduce the controversy to only two alternatives.

Maintaining Perspective

In examining the creation-evolution controversy, it may become clear that one of the reasons for the controversy is the particular emphasis of the Christian faith.The primary concern of the Bible is not with issues of science. It is concerned with describing our relationship with God and delineating how we should live with regard to that relationship. The Bible mainly deals with moral issues, not scientific ones. Therefore, in the Bible the Christian will find much more certainty about the spiritual and moral aspects of life than about matters of science. The disciplines of science and religion have separate purposes and therefore will differ in their treatment of the same subject.

Christian theism does not stand or fall on whether one can prove that the Bible is right about origins. The Christian faith is concerned with the present state of the human race, not with what the race was like a thousand or ten thousand years ago. As Louise Young says, "Science is one way of getting at the truth—some people believe, the only way—but the truth which it reveals is of a tentative nature. It does not provide ultimate or absolute truth such as that claimed by religious revelations or philosophical systems."[3] Therefore, we should not expect science to do something for which it is not designed nor should we expect the Bible to reveal truths to us which God has chosen to keep hidden.

Guidelines for Dealing with Bible-Science Conflicts

The following guidelines will help you cut through the confusion surrounding difficult controversies and help you learn how to make up your own mind about the issues.

1. *Define the problem.* This will give you a direction for research into the issue. It will also help you to know what the answer will look like when you find it.

2. *Define your terms.* Everyone should know what is meant when certain key words are used. This is especially

important in the creation-evolution conflict because many of the terms which are frequently used have several meanings. For example, the terms *creationist* and *evolutionist* can apply to several different views.

3. *Recognize the assumptions and biases involved.* "Each, science and religion, has its own basic assumptions and procedures."[4] You should be able to discover what assumptions you or others hold. You should also realize that once a person adopts and uses a particular theory or model, it becomes very difficult to think within the framework of other theories and models.

4. *Distinguish science from pseudoscience and nonscience.* Knowing the scientific method well is a big help. Because they are human, scientists sometimes allow unrecognized and nonscientific value judgments to slip into their investigations. Science has definite limits which must be observed.

5. *Listen to all sides of the argument.* Even though you may strongly disagree, you should listen to your opponents. You should also be willing to admit truth wherever it is found. One position is seldom completely right and the other completely wrong. Besides, you want them to give you a fair hearing, too, don't you?

6. *Distinguish between evidence and interpretation.* Although both are subject to personal bias, interpretation is far more susceptible.

7. *Recognize when there is insufficient evidence.* Dr. Frank Crane admits that "quite often a scientist speaks before all the evidence is in. He has considered one or some possibilities, but not all. Some Christians follow the same illogical pattern of speaking dogmatically before all the evidence is in. Much evidence is yet to be uncovered by both Christians and scientists."[5] Premature answers are useful only if used as hypotheses to be tested.

8. *Do not insist upon complete agreement between un-*

changing Scripture and changing science. It will inevitably lead to problems in the future.

9. *Pray.* Pray for the wisdom to identify truth and the discernment to recognize mysteries.

If you use these guidelines and familiarize yourself with as much of the relevant material as possible, you may be able to draw your own conclusions with regard to creation-evolution and similar issues.

The greatest danger in any emotional controversy is that you will determine your position on an issue without first considering all sides. Kerkut affirms:

> It would seem a good principle to encourage the study of "scientific heresies." There is always the danger that a reader might be seduced by one of these heresies but the danger is neither as great nor as serious as the danger of having scientists brought up in a type of mental strait-jacket or of taking them so quickly through a subject that they have no time to analyse and digest the material they have "studied."[6]

In some cases, scientists have recognized that they have been guilty of an ignorant dogmatism:

> What we have all accepted as the Whole truth, turns out with mild inspection, to be rather far from it. Apparently, if one reads the original papers instead of relying on some superficial remarks in a textbook, the affinities become extremely clouded indeed. We have all been telling our students for years not to accept any statement on its face value but to examine the evidence, and therefore, it is rather a shock to discover we have failed to follow our own sound advice.[7]

In a time when some Christians call us to affirm their theory as the only one possible for a biblical Christian, it is important to refuse to be rushed into decisions. If the argument of this book is sound, we do not have enough evidence to say that one version of origins should with absolute certainty be espoused over another. We should listen to others, Christian and non-Christian alike. We should be wary of those who make one theory the test of whether one is truly a biblical Christian. It is important to know how to discuss the creation-evolution issue intelligently, but it is not the most important issue in a Christian's life.

Notes

Chapter 1

[1]David J. Merrell, *Evolution and Genetics: The Modern Theory of Evolution* (New York: Holt, Rinehart and Winston, 1962), pp. 14-15.
[2]Thomas Kuhn, *The Copernican Revolution* (Cambridge, Mass.: Harvard Univ. Press, 1957), p. 199.
[3]Sir William Cecil Dampier, *A History of Science and Its Relations with Philosophy and Religion* (New York: Macmillan, 1944), pp. 71-72.
[4]Gerald James Stine, *Biosocial Genetics: Human Heredity and Social Issues* (New York: Macmillan, 1977), p. 37.
[5]Stine, p. 56.
[6]Ibid., pp. 57-58.
[7]Dorothy Nelkin, "The Science-Textbook Controversies," *Scientific American*, 234 (1976), 33-39.
[8]Ibid.
[9]"Law against Teaching Evolution Overturned," *Tulsa Tribune*, 21 Dec. 1970.
[10]Walter G. Peter III, "Fundamentalist Scientists Oppose Darwinian Evolution," *Bioscience*, 20 (1970), 1067-69.
[11]Richard A. Dodge, "Divine Creation: A Theory?" *AIBS Education Review*, 2 (1973), 29-30.
[12]Elwood B. Ehrle, "California's Anti-evolution Ruling," *Bioscience*, 20 (1970), 291; and Peter, pp. 1067-69.
[13]Dodge, pp. 29-30.
[14]Ehrle, p. 291.
[15]Michael D. Byer, "Letters to the Editor," *Bioscience*, 20 (1970), 642.
[16]Ellen Weaver quoted in Ehrle, p. 291.
[17]Karl D. Fezer, "Letters to the Editor," *Bioscience*, 20 (1970), 641-42.
[18]D. W. Bullock, "Letters to the Editor," *Bioscience*, 20 (1970), 524.
[19]Carl Gans, "Letters to the Editor," *Bioscience*, 20 (1970), 844.
[20]Janice M. Glime, "Letters to the Editor," *Bioscience*, 20 (1970), 642.
[21]J. Wanless Southwick, "Letters to the Editor," *Bioscience*, 20 (1970), 641. Emery F. Swan, "Letters to the Editor," *Bioscience*, 20 (1970), 640.
[22]John C. Hendrix, "Letters to the Editor," *Bioscience*, 20 (1970), 524.
[23]Robert Tricardo, "Letters to the Editor," *Bioscience*, 20 (1970), 992-93.
[24]Walter C. Kraatz, "Letters to the Editor," *Bioscience*, 20 (1970), 641.
[25]Wayne Frair, "Letters to the Editor," *Bioscience*, 20 (1970), 642.
[26]"The Devil's Advocate," *American Biology Teacher*, 32 (1970), p. 495.
[27]Duane T. Gish, "A Challenge to Neo-Darwinism," *American Biology Teacher*, 32 (1970), 497.
[28]Thomas J. Cleaver, "Letters to the Editor," *American Biology Teacher*, 33 (1971), 300.
[29]"Judge Rules against Text on Creation," *Tulsa Tribune*, 13 April 1977.
[30]William Willoughby, "Washington Perspective" rpt. in *Acts & Facts*, 1, No. 3 (1972), 6.
[31]See William Bevan, "Two Cooks for the Same Kitchen?" *Science*, 177 (1972), 1155; and Nicholas Wade, "Creationists and Evolutionists: Confrontation in California," *Science*, 178 (1972), 724-29.
[32]Rpt. in *American Biology Teacher*, 35 (1973), 15-224. See also "Creationists Reply to Cory Article," *American Biology Teacher*, 36 (1974), 48-50.

[33]Jerry P. Lightner, ed., *A Compendium of Information on the Theory of Evolution and the Evolution-Creation Controversy* (Reston, Va.: NABT, 1977).
[34]Bruce Wallace, "Teaching of Evolution," a letter to the members of the Society for the Study of Evolution, 1973.
[35]"A Statement Affirming Evolution as a Principle of Science," *The Humanist*, 37 (1977).
[36]Stine, p. 59.
[37]"California School Decision: Victory or Defeat?" *Acts & Facts*, 2, No. 1, 1-3.
[38]Nelkin, p. 34.

Chapter 2

[1]J. Edwin Orr, *100 Questions about God* (Glendale, Calif.: G/L Publications, 1966), p. 64.
[2]G. A. Kerkut, *Implications of Evolution* (New York: Pergamon, 1960), p. viii.
[3]Lois Anne Nagy and Bartholemew Nagy, "Interdisciplinary Search for Early Life Forms and for the Beginning of Life on Earth," *Interdisciplinary Science Reviews*, 1 (1976), 291.
[4]Ernst Mayr, *Populations, Species, and Evolution* (Cambridge, Mass.: Harvard Univ. Press, Belknap Press, 1970), p. 10.
[5]J. D. Thomas, *The Doctrine of Evolution and the Antiquity of Man* (Abilene: Biblical Research Press, 1963), p. 7.
[6]John R. W. Stott, *Your Mind Matters* (Downers Grove, Ill.: InterVarsity Press, 1972), p. 21.
[7]Jay M. Savage, *Evolution, Second Edition* (New York: Holt, Rinehart and Winston, 1969), pp. v-vi.
[8]G. Ledyard Stebbins, Jr., *Variation and Evolution in Plants* (New York: Columbia Univ. Press, 1950), p. x. Stebbins and some other evolutionists later enlarged the scope of microevolution to include the formation of species through hybridization and subsequent polyploidy. (If the number of chromosomes in the hybrid is doubled, the new species can interbreed, thereby confirming it as a legitimate species.) For our purposes in this volume speciation through hybridization will be considered an example of microevolution (see Figure 6). The formation of two new species from one old one will be treated as an aspect of macroevolution. These problems in defining *microevolution* and *macroevolution* are the result of disagreement over the exact definition of species.
[9]Kerkut, p. 157.
[10]Richard A. Walker, Thomas R. Mertens and Jon R. Hendrix, "Clarifying the Creation-Evolution Issue with Biology Students," *American Biology Teacher*, 39 (1977), 50.
[11]Savage, p. x.
[12]Paul E. Little, *Know Why You Believe* (Downers Grove, Ill.: InterVarsity Press, 1968), p. 79.
[13]Mayr, p. 1.
[14]Louise B. Young, ed., *Evolution of Man* (New York: Oxford Univ. Press, 1970), p. 67.
[15]Kerkut, pp. 6-7.
[16]Ibid., pp. 155-56.
[17]Thomas, pp. 7-8.
[18]Nagy and Nagy, p. 307.
[19]Little, p. 70.
[20]John Tyler Bonner, "Perspectives," *American Scientist*, 49 (1961), 242.

[21]Kerkut, p. vii.
[22]Little, p. 50.
[23]Orr, p. 83.
[24]Bonner, p. 242.
[25]W. E. Le Gros Clark, *The Fossil Evidence of Human Evolution* (Chicago: Univ. of Chicago Press, 1955), p. 24.

Chapter 3

[1]Paul B. Weisz and Richard N. Keogh, *Elements of Biology*, 4th ed. (New York: McGraw-Hill, 1977), pp. 7-9.
[2]"Professors Test Theory," *Capper's Weekly*, 9 Nov. 1976.
[3]Claude A. Villee, *Biology*, 7th ed. (Philadelphia: Saunders, 1977), p. 6.
[4]Paul Milvy, "Getting to the Heart," *Runners World*, 12 (1977), 27.
[5]Joseph Wood Krutch, "How Right Was Darwin?" in Young, pp. 113-16.
[6]Bernard G. Campbell, *Human Evolution: An Introduction to Man's Adaptation*, 2nd ed. (Chicago: Aldine, 1974), preface.
[7]Weisz and Keogh, pp. 7-9.
[8]Stephen L. Talbot, "Research Communications Network Newsletter," No. 2 (1977), 2.
[9]Bruce Wallace, *Chromosomes, Giant Molecules, and Evolution* (New York: Norton, 1966), pp. 4-5.
[10]Don G. Stafford, John W. Renner and John J. Rusch, *The Physical Sciences: An Inquiry and Investigation* (Beverly Hills, Calif.: Glencoe, 1977), p. 41.
[11]W. I. B. Beveridge, "The Nature of Scientific Theory" in Young, p. 59.
[12]Campbell, preface.
[13]Theodore Delevoryas, *Plant Diversification* (New York: Holt, Rinehart and Winston, 1966), p. vi.
[14]Savage, pp. v-vi.
[15]Herbert Ross, "The Logical Bases of Biological Investigation," *Bioscience*, 16 (1966), 16.
[16]Robert Sharvy, *Logic: An Outline* (Totowa, N. J.: Littlefield, 1970), pp. 106-07.
[17]J. B. Conant, *Science and Common Sense* (New Haven, Conn.: Yale Univ. Press, 1961), p. 44.
[18]See Lawrence R. Cory, "Creationism and the Scientific Method," *American Biology Teacher*, 35 (1973), 223-24.
[19]Ross, p. 15.
[20]Loren Eiseley, *Darwin's Century: Evolution and the Men Who Discovered It* (Garden City, N.Y.: Doubleday, Anchor, 1961), p. 62.
[21]Cordelia E. Barber, "Fossils and their Occurrence" in *Evolution and Christian Thought Today*, ed. R. L. Mixter (Grand Rapids, Mich.: Eerdmans, 1959), p. 153.
[22]Weisz and Keogh, p. 10.
[23]Warren W. Weaver, "Who Speaks for Whom and for What?" *Science*, 119 (1954), 3A.
[24]Morris Goran, *Science and Anti-science* (Ann Arbor, Mich.: Ann Arbor Science, 1974), p. 77.
[25]E. Peter Volpe, *Understanding Evolution* (Dubuque: Wm. C. Brown, 1967), p. 15.
[26]G. Ledyard Stebbins, Jr., *Flowering Plants: Evolution above the Species Level* (Cambridge, Mass.: Harvard Univ. Press, 1974), pp. viii-ix.
[27]John N. Moore, *Should Evolution Be Taught?* (San Diego: Creation-Life, 1974), p. 8.
[28]George G. Simpson, *The Meaning of Evolution*, rev. ed. (New Haven, Conn.: Yale Univ. Press, 1967), p. 279.

Chapter 4

[1]Theodore H. Eaton, *Evolution* (New York: Norton, 1970), p. 3.
[2]Julian Huxley quoted in "Panel Two: The Evolution of Life" in *Evolution after Darwin*, 3, ed. Sol Tax (Chicago: Univ. of Chicago Press, 1960), p. 111.
[3]Theodosius Dobzhansky, "Evolution" in Young, p. 58.
[4]Thomas C. Emmel, *Worlds within Worlds: An Introduction to Biology* (New York: Harcourt, 1977), p. 210.
[5]George H. M. Lawrence, *Taxonomy of Vascular Plants* (New York: Macmillan, 1951), p. 50.
[6]Mayr, p. 3.
[7]Andrew Dickson White, *A History of the Warfare of Science with Theology in Christendom*, (New York: Dover, 1896), Vol. I.
[8]Ralph W. Dexter, "Historical Aspects of Louis Agassiz's Lectures on the Nature of Species," *Bios* 48 (1977), pp. 12-19.
[9]Barber, p. 153.
[10]Eaton, p. 18.
[11]Mayr, p. 12.
[12]Lawrence, p. 50.
[13]Eaton, p. 20.
[14]R. B. Goldschmidt, "Evolution, as Viewed by One Geneticist," *American Scientist*, 40 (1952), p. 96.
[15]Young, p. 136.
[16]Mayr, p. 6.
[17]Ludwig von Bertalanffy, "Evolution" in Young, p. 120.
[18]Moore, pp. 26-27.
[19]Henry M. Morris, *The Twilight of Evolution* (Grand Rapids, Mich.: Baker Book House, 1963), p. 67.
[20]Emmel, pp. 208-16.
[21]Eaton, p. 117.
[22]Ludwig von Bertalanffy, "Evolution" in Young, pp. 116-17.
[23]Ibid.
[24]See Moore.
[25]Russell L. Mixter, "A Wheaton College View of Creation and Evolution" mimeographed by Wheaton College, Wheaton, Ill., 1957.
[26]Coleman J. Goin and Olive B. Goin, *Journey onto Land* (New York: Macmillan, 1974), p. 101.

Chapter 5

[1]Bonner, p. 242.
[2]Paul W. Hodge, *Concepts of Contemporary Astronomy* (New York: McGraw-Hill, 1974), p. 531.
[3]Ibid., pp. 101-08.
[4]George K. Schweitzer, "The Origin of the Universe" in Mixter, *Evolution and Christian Thought Today*, p. 51.
[5]Emmel, p. 228.
[6]Charles Darwin, *The Origin of Species by Means of Natural Selection* (New York: Modern Library, 1872), p. 374.
[7]Nagy and Nagy, pp. 292-302.
[8]Karl von Naegeli quoted in Garrett Hardin, *Biology: Its Principles and Implications* (San Francisco: W. H. Freeman, 1961), p. 213.
[9]Emmel, p. 230.

[10]Harry J. Fuller and Oswald Tippo, *College Botany*, rev. ed. (New York: Holt, Rinehart and Winston, 1961), p. 25.
[11]Carl O. Dunbar, *Historical Geology*, 2nd ed. (New York: Wiley, 1961), p. 47.
[12]Barber, pp. 150-51.
[13]Stebbins, *Flowering Plants*, p. viii.
[14]Goin and Goin, p. 30.
[15]George G. Simpson, *Tempo and Mode in Evolution* (New York: Columbia Univ. Press, 1944), p. 99.
[16]George G. Simpson, *The Meaning of Evolution* (New Haven, Conn.: Yale Univ. Press, 1949), p. 31. See also Simpson, *The Meaning of Evolution*, rev. ed. (1967), p. 33.
[17]George G. Simpson, *The Major Features of Evolution* (New York: Columbia Univ. Press, 1953), p. 36.
[18]Everett C. Olson, "The Role of Paleontology in the Formation of Evolutionary Thought," *Bioscience*, 16 (1966), 39.
[19]Daniel Axelrod, "Early Cambrian Marine Fauna," *Science*, 128 (1958), 7-9.
[20]Nagy and Nagy, p. 296.
[21]Bonner, p. 240.
[22]Barber, pp. 150-51.
[23]Dwight D. Davis, "Comparative Anatomy and Evolution of Vertebrates" in *Genetics, Paleontology, and Evolution*, ed. G. L. Jepson, G. G. Simpson and Ernst Mayr (Princeton, N. J.: Princeton Univ. Press, 1949), p. 77.
[24]Daniel I. Axelrod, "Evolution of the Psylophyte Paleoflora," *Evolution*, 13 (1959), 247.
[25]John N. Moore, "On Chromosomes, Mutations, and Phylogeny" read before the Society for the Study of Evolution, 138th annual meeting, 27 Dec. 1971, Philadelphia.
[26]Duane T. Gish, "Petroleum in Minutes, Coal in Hours," *Acts & Facts*, 1, No. 4 (1972), 1-5.
[27]Duane T. Gish, "Speculations by a Scientist—Radiochronological Clocks in Shambles?" *Acts & Facts*, 2, No. 2 (1973), 4.
[28]Harold S. Slusher, "The Age of the Solar System," pt. 1, *Acts & Facts*, 2 (1973), 1-3.
[29]Eaton, p. 197.
[30]Ian Tattersall and Niles Eldridge, "Fact, Theory, and Fantasy in Human Paleontology," *American Scientist*, 65 (1977), 204-10.
[31]Young, p. 4.
[32]Ross, p. 16.
[33]Kerkut, p. 154.
[34]Bonner, pp. 243-44.
[35]Ibid., p. 242.
[36]Savage, p. 119.
[37]Ibid., p. 131.
[38]E. J. H. Corner quoted in *Contemporary Botanical Thought*, ed. Anna M. Cobley and L. S. Cobley (Chicago: Quadrangle Books, 1961), p. 97.

Chapter 6

[1]Donald England, *A Christian View of Origins* (Grand Rapids, Mich.: Baker Book House, 1972), pp. 116-17.
[2]Orr, pp. 85-97.
[3]Ibid., p. 69.
[4]David D. Riegle, *Creation or Evolution?* (Grand Rapids, Mich.: Zondervan,

1971), p. 19. Riegle cites the following passages: Mt. 19:4-7; Rom. 5:12-19; 1 Cor. 15:45; 1 Tim. 2:13-14; see also Francis Schaeffer, *No Final Conflict* (Downers Grove, Ill.: InterVarsity Press, 1975), p. 14, for comments on the biblical record.

[5]R. C. Sproul, *Knowing Scripture* (Downers Grove, Ill.: InterVarsity Press, 1977), p. 49.

[6]The following are also recommended for study: J. Stafford Wright, *Interpreting the Bible* (Downers Grove, Ill.: InterVarsity Press, 1955); and Bernard Ramm, *Protestant Biblical Interpretation* (Grand Rapids, Mich.: Baker Book House, n.d.); and T. Norton Sterrett, *How to Understand Your Bible* (Downers Grove, Ill.: InterVarsity Press, 1974).

[7]Riegle, p. 52.

[8]White.

[9]Schaeffer, pp. 25, 37-43.

[10]Paul Zimmerman, "The Age of the Earth" in *Darwin, Creation and Evolution* (St. Louis: Concordia, 1959), p. 161.

[11]John C. Whitcomb and Henry M. Morris, *The Genesis Flood* (Philadelphia: Presby. & Reformed, 1965); Harold W. Clark, *Fossils, Flood, and Fire* (Escondido, Calif.: Outdoor Pict., 1968); Henry M. Morris, ed., *Scientific Creationism* (San Diego: Creation-Life, 1974); and David C. Watson, *The Great Brain Robbery* (Chicago: Moody, 1976).

[12]R. J. Wiseman, *Creation Revealed in Six Days* (London: Marshall, Morgan, & Scott, 1949).

[13]Martin Gardner, *Fads and Fallacies in the Name of Science* (New York: Dover, 1957), pp. 124-26.

[14]Additional discussions of theistic evolution can be found in the following literature: Richard Bube, ed., *The Encounter between Science and Christianity* (Grand Rapids, Mich.: Eerdmans, 1968); Morris, *Scientific Creationism;* Watson, *The Great Brain Robbery;* Schaeffer, *No Final Conflict;* and Bernard Ramm, *The Christian View of Science and Scripture* (Grand Rapids, Mich.: Eerdmans, 1954).

[15]Peter Stoner, *Science Speaks* (Chicago: Moody Press, 1952); Ramm, *The Christian View of Science and Scripture;* Mixter, *Evolution and Christian Thought Today;* and Morris, *Scientific Creationism.*

Chapter 7

[1]Because of the complexity of the creation-evolution issue, you may well run into instructors who do not recognize that there is any other side to the issue but theirs. If your instructors require that answers to their test questions conform to their own personal beliefs or to a biased textbook, give them the answers they request. Consider the test questions as the instructor's attempts to learn if you know the material presented to you (which may differ from the facts). You can hold your own personal beliefs, which may be quite different from those of your instructor, but you need not constantly try to change your teacher's mind.

[2]Walter R. Hearn and R. A. Hendry, "The Origin of Life" in Mixter, *Evolution and Christian Thought Today,* pp. 67-69.

[3]Young, p. 4.

[4]Goran, p. 20.

[5]Frank A. Crane, "Evidences of God in Plant Life" in Hefly, p. 31.

[6]Kerkut, p. 157.

[7]Bonner, p. 240.